DISCARD

Evolution and Religion

New
Dialogues in
Philosophy

A series in dialogue form, explicating foundational problems
in the philosophy of existence, knowledge, and value

Series Editor
Professor Dale Jacquette, Senior Professorial Chair in Theoretical Philosophy, University of Bern, Switzerland

In the tradition of Plato, Berkeley, Hume, and other great philosophical dramatists, Rowman & Littlefield presents an exciting new series of philosophical dialogues. This innovative series has been conceived to encourage a deeper understanding of philosophy through the literary device of lively argument in scripted dialogues, a pedagogic method that is proven effective in helping students to understand challenging concepts while demonstrating the merits and shortcomings of philosophical positions displaying a wide variety of structure and content. Each volume is compact and affordable, written by a respected scholar whose expertise informs each dialogue, and presents a range of positions through its characters' voices that will resonate with students' interests while encouraging them to engage in philosophical dialogue themselves.

Titles
J. Kellenberger, *Moral Relativism: A Dialogue* (2008)
Michael Ruse, *Evolution and Religion: A Dialogue* (2008)
Charles Taliaferro, *Dialogues About God* (2008)

Forthcoming titles
Bradley Dowden, *The Metaphysics of Time: A Dialogue*
Dale Jacquette, *Dialogues on the Ethics of Capital Punishment*
Michael Krausz, *Relativism: A Dialogue*
Dan Lloyd, *Ghosts in the Machine: A Dialogue*
Brian Orend, *On War: A Dialogue*

Evolution and Religion

A *Dialogue*

Michael Ruse

ROWMAN & LITTLEFIELD PUBLISHERS, INC.
Lanham • Boulder • New York • Toronto • Plymouth, UK

ROWMAN & LITTLEFIELD PUBLISHERS, INC.

Published in the United States of America
by Rowman & Littlefield Publishers, Inc.
A wholly owned subsidiary of The Rowman & Littlefield Publishing Group, Inc.
4501 Forbes Boulevard, Suite 200, Lanham, Maryland 20706
www.rowmanlittlefield.com

Estover Road
Plymouth PL6 7PY
United Kingdom

Copyright © 2008 by Rowman & Littlefield Publishers, Inc.

All rights reserved. No part of this publication may be reproduced, stored in a
retrieval system, or transmitted in any form or by any means, electronic, mechanical,
photocopying, recording, or otherwise, without the prior permission of the publisher.

British Library Cataloguing in Publication Information Available

Library of Congress Cataloging-in-Publication Data

Ruse, Michael.
  Evolution and religion : a dialogue / Michael Ruse.
    p. cm.
  Includes bibliographical references.
  ISBN-13: 978-0-7425-5906-6 (cloth : alk. paper)
  ISBN-10: 0-7425-5906-8 (cloth : alk. paper)
  1. Religion and science. I. Title.
  BL240.3.R87 2008
  201'.65—dc22
                                2007047304

Printed in the United States of America

∞™ The paper used in this publication meets the minimum requirements of
American National Standard for Information Sciences—Permanence of Paper
for Printed Library Materials, ANSI/NISO Z39.48-1992.

~

Contents

~

Preface

This book is written for my students and, truly, for everyone else's students also. So in a way, it is dedicated to all of them. But, I would like to thank my wife, Lizzie Matthews, my children, Emily Constance Ruse, Oliver John Fentiman Ruse, and Edward George Redvers Ruse, as well as my in-laws, Harold Matthews and Nancy Wallace, for helping with the names. Charles Dickens's stories gave me the idea for the name of the person who most closely represents my position. Thanks also to Dale Jacquette for a great idea.

I worried a little about whether this dialog should be called *Science and Religion* or, more specifically, *Evolution and Religion*. In the end, I decided that the second title was much more appropriate. But I wanted to feel free to range somewhat, so I framed the discussion as being about science and religion and then let the expertise and interests of my participants decide the actual content. This way, the first chapter could be a general introduction to the science-religion interface. Later, although this impinges on physics, I particularly wanted to include a discussion of the Anthropic Principle. It seems to me that now the physicists are trying to revive the argument from design for God's existence, something that evolutionary biologists have been discussing rather critically for over a hundred years. So the discussion there is pertinent to the general theme. Also, as you will see towards the end, I wanted to bring in some general issues about humankind. Here too, however, evolution is not far away.

~

CHAPTER ONE

Program One: Options

Good evening. My name is Redvers Fentiman and tonight, and for the next four Tuesdays, I invite you to join our popular, public-television-sponsored series of *Eternal Questions*. Our theme for this new program is "Science and Religion: Who Is Winning?" We have a terrific lineup of guests here to discuss this issue, and I am sure it will be a fascinating experience for us all. Introducing the discussants, on my far left, we have Professor David Davies, head of the Department of Evolutionary Biology at the Massachusetts State Institute of Science and the author of several books, most recently *Genes, God, and Gollum*. Then next to him, on my immediate left, is Martin Rudge, historian and philosopher of science at Robert Boyle, the well-known liberal arts college in Minnesota.

Then, crossing over, on my immediate right, is the Reverend Emily Matthews, Episcopalian priest and adjunct professor of pastoral counseling at Wycliffe College, here in this city. And finally, on my far right—I'm not yet talking about his politics or theology—is the Reverend Harold Wallace, head pastor of a very large Southern Baptist church, Rollingbrooke Stones, in Atlanta. I'm right, am I not, Dr. Wallace, in saying your church is very large?

Wallace: Please call me Hal. Yes, we have over seven thousand members, and twenty-three pastors, including me. We started with just a room and ten members twenty years ago and now we have an annual budget of $14 million.

Rudge: I obviously went into the wrong business. I got a grant last year for $5,000 and I felt lucky.

Davies: Yes, but you're not into the God business. You may be offering education, but you're not offering eternal salvation. You may flunk kids, but you don't burn them all to hell for eternity. You should get people really scared about the future and then watch how the money rolls in. In my business, all you have to do is find another disease and some sports figure is up there on television raising money for you. It's the same with religion. Look at who's the best-dressed person on this panel!

Matthews: Well, that's certainly not me! My partner had to bully me to not just wear a sweater and blue jeans. But you know, David, although as a priest I am deeply ashamed of the behavior of some of my fellow ministers—remember Jim and Tammy Faye Bakker—you're wrong tarring us all with the same brush. How would you like it if we said that all scientists are just cold materialists, who sell their souls to the highest bidder? Werner von Braun, the rocket scientist, who didn't care if he worked for Hitler or for the Americans, is hardly a model for all scientists. Most of us who are ordained ministers are very, very far from being rich. Like most everyone else, we just want to make a living so we can live decently.

In my parish, by far the bulk of the collection goes to service. We run a hostel and kitchen over in the ward—you know, where the strip clubs are—and we outreach to Ethiopia. I'm not saying that Professor Rudge didn't deserve his money, but with all due respect—and I speak now as someone who is a bit in the same business herself—it is not as if sitting on your bottom in a comfortable office, doing the kind of thing you like, is of the same value as curing a person sick with AIDS or offering a helping hand to the lonely and hurt and helpless, here in our ghettos or there in Africa.

Davies: I hate to be rude right from the beginning, or perhaps it is just as well to be rude right from the beginning, but, Reverend—

Matthews: Call me Emily.

Davies: Okay, Emily, but right from the beginning I have to say that not only do I not buy into that helping-hand stuff—or at least I do, but not in the way that you would do it—but I think that everything and anything to do with religion is not only a waste of time, but positively dangerous. If it just helped people feel better, then why not? I wouldn't ban pot and so why should I ban religion? But the thing is that pot is not dangerous, except possibly marginally to the smoker. But religion is dangerous: Northern Ireland, Israel, Iran, Pakistan, and on and on. You name it. Where there's conflict, there's religion. As the Nobel Prize-winning, physicist Steven Weinberg has said, "Good people will do good things, and bad people will do bad things. But for good people to do bad things—that takes religion." And before you start in about soup kitchens and the like, I'd like to know about the strings. The hymns and the prayers that you have to hear before you can get a hamburger and fries. Do you have to light a candle before you can get ketchup?

Wallace: You know, what you and Weinberg say about religion is not just false, but rather offensive. I reject absolutely the idea that my people are doing bad things because of their faith. When Hurricane Katrina hit New Orleans, where were the physicists? Where were the philosophers? I don't know. I didn't see any, and if they had been there, I would have seen them. Teams from my church spent two months working there, operating a kitchen, finding clothing and bedding, opening a day-care center for the kids, and much more—much more than the government, I can tell you.

Fentiman: I can see that we have strong opinions here tonight, and I am sure that we all want to express them forcefully. But, we are not here just to talk about religion, but about religion and science. So, let's focus ourselves on that. We are in the *Eternal Questions* series, and our topic is science and religion, and the conflict. Or not, as you might want to say.

Davies: But isn't that the whole point? You think it's an eternal question. I would challenge that right from the beginning. Why is it an eternal question? It was a question, but it's been answered. Science is right and religion is wrong. Or more precisely, religion is wrong and dangerous and science is on the road to truth. In fact, I am very keen on Karl Popper. He said that you could never ever prove anything true, but you could prove it false. So I would say that science may never be true, but it's the best way that we have of not being false. Discussion over.

Fentiman: Well, I hope not because we have four more hours after this one to discuss the issues, and it is costing our sponsors a lot of cash to put this on. So, I think we should labor on a bit because I doubt that everyone on the panel agrees with you. What I'd like to do, since I am the host, is structure the discussion a bit. Martin Rudge, we haven't heard much from you. Just sitting on your bottom contemplating that next book, I guess! Anyway, I'd like you to kick off the debate. Perhaps you could just give us a bit of the history of things—science and religion—and spell out the options for today.

Rudge: Fair enough. Stop me if I start to sound a bit too professorial. Although, let me say that I don't take offense at what Emily said—not that I think it was meant very offensively—and indeed, I do worry about what an academic like me can contribute to society's well-being, let alone the horrendous problems across the world. One thing, it may not be much, is to take part in TV programs like this! We are talking about really important issues and, as we are going to see, these issues have serious implications for education, social policy, and a lot more—who should or should not go to prison, for example.

In a funny sort of way, I kind of agree with David Davies that the science-religion relationship is not an eternal question. But, I say that from completely the opposite end from him. He thinks that it is over. I'm not so sure about that. But, I do think it had a beginning—up to the end of the Middle Ages, that is, until the Reformation in the sixteenth century, as well as what we call the Scientific Revolution—Copernicus to Newton—by and large, science and religion went hand in glove. To a large degree, the only people who could read and write were the clergy, and they were the only ones with the leisure to look into scientific questions. The Catholic Church was very pro science. Don't forget that Copernicus, the guy who put the sun at the center of the universe, was a minor cleric. And he died in good standing. No excommunication for him!

Davies: Yes, but don't forget Galileo—a hundred years later, made to get down on his knees and say the earth does not move and the sun is not at the center.

Rudge: I'm not forgetting Galileo. I'm just getting to him. Two things were very important for splitting science and religion. Obviously, the rise of the Protestants was one thing. It simply isn't true that back then they were all crude biblical literalists like we have in America today, but people like Luther and Calvin did take the Bible very seriously, and there is no doubt but that the claims in the holy book often clashed with the new science, and religious people felt tense. The science itself also contributed to the tension. Copernicus knocked the earth from the center of the universe. Up to then, everyone agreed with the Greek philosopher Aristotle that the earth had to be the center, because that is why heavy things fell down and not up. But with Aristotle pushed sideways in science, this put pressure on the Catholics, because their theology—thanks to Saint Thomas Aquinas in the thirteenth century—used a lot of Aristotle. The idea that while the physical appearance stays the same, the bread and the wine can really turn into the body and blood of Christ—transubstantiation—is very Aristotelian. So as the science went, the religion felt pressure.

Funnily enough, scholars today think the Galileo episode was a bit of an anomaly. No one ever lost their life because of Copernicus. It used to be said that Giordano Bruno was burned at the stake in 1600 for being a Copernican. But that is not really so. He got barbecued because he claimed that there is life on other planets, and that was heretical. Galileo was a bit unlucky. He was made to recant in 1632. By then the Catholic Church was going full blast on the Counter-Reformation, and was trying to outdo the Protestants on being tough on claims that might threaten the Bible. But, Galileo's real sin was that he published in Italian and so made the ideas readily available and hence dangerous. Also, he was a bit of a pain in the butt. He really did portray his opponents as fools and knaves. He should not have been condemned, but still.

Wallace: Yes, but this is all very well. What about Darwin? There was someone who went absolutely against the Bible and, hence, against the Christian religion. They can't both be right. It's Jesus Christ or Charles Darwin. Take your choice. I know where I stand—with Pat Robertson: "God is tolerant and loving, but we can't keep sticking our finger in his eye forever."

Fentiman: I wonder if we can just leave Darwin out of the discussion for the moment, at least the details. I am hoping that we can get to Darwin next week and have a full discussion of him and his theory then. Just so you don't get too professorial, Martin, why don't you wrap things up a bit. Where do we stand now?

Rudge: The most popular classification, due to a man called Ian Barbour, a retired professor from Carleton College, a neighbor college of mine in Minnesota, has four divisions or options. First, there is the *warfare* division or metaphor. Here, science and religion are seen as being in a nonstop battle. If you are for Genesis, then you are against evolution. End of debate, or at least end of any attempt at finding harmony. The second division goes the other way completely. It is the *independence* position. It says that science and religion are simply talking about different things. They are chalk and cheese. They do not overlap and, hence, they could not conflict if they wanted to. This is often known as "neo-orthodoxy," because it is inspired by the great Swiss theologian of the first half of the twentieth century, Karl Barth.

The other two divisions are reasonably close together. One speaks of harmony or *dialogue*. Here, science and religion are considered to be separate, but they can interact fruitfully. I suspect a lot of Catholics, including the present pope, Benedict XVI, feel drawn to this position. The final division is one of *integration*. It says that all knowledge is one, and ultimately science and religion will be seen as part of the whole. Followers of Alfred North Whitehead, the English logician who went to Harvard in the first part of the last century, are often put in this group. These are the so-called Process Philosophers.

Fentiman: That's a good start. Now, let's get the others back into the conversation. Starting out in left field with David Davies, where do you stand?

Davies: You all know where I stand. I am a conflict man all the way. Except, I think the conflict is over, because science has won hands down. But let's not just say that. Let's see why that's the case. First, although Martin has been glossing over this a bit, history shows this solidly. Back in medieval times, the church and science might have been buddy-buddy, but it was on the church's terms all of the way. Richard Dawkins, the English biologist, has rightly spoken of the poky little universe. You had a flat earth, with nothing much beyond Europe and the Near East, as we now know it, and the sun and the stars were just above, with God and the angels out there beyond and a lot of fire and the devil down below. There was a short history to the earth—I won't go into that now because we are going to talk about this topic next time—but most important of all, there were miracles everywhere. Not just the saints curing sick people, but old women putting curses on young women so they couldn't have babies, when they were not hopping on broomsticks to go off and have fun with the devil, with all sorts of omens appearing in battles telling you what was going to happen, with superstition rampant—potions, zodiac signs, mysterious fires, you name it.

What modern science has done is to wipe all of this away, but only with religion kicking and screaming the whole time. Darwin's great supporter Thomas Henry Huxley once said—and I have brought the quote along so I get it right—"Extinguished theologians lie about the cradle of every science as the strangled snakes beside that of Hercules; and history records that whenever science and orthodoxy have been fairly opposed, the latter has been forced to retire from the lists, bleeding and crushed if not annihilated; scotched, if not slain." He was absolutely right. I don't care who Copernicus was. It was what he thought and said that counts. He pushed the earth sideways, and, much more important, he made the universe vastly bigger—four hundred thousand times bigger, in fact. This came about simply because if Copernicus is right and the earth is moving around the sun, this motion should affect the way in which you see the stars. If you walk from front to back of an automobile, now you see the hood and now you see the trunk. The point is that no one could see this effect with the stars—it's known as "stellar parallax"—and the only way you can explain its absence is by supposing that the stars are too far away to spot it. If you put an automobile between the goalposts at one end of the football field, and then go down to the other end and walk between the posts, you will see the car in much the same way left or right. It's too far away to spot the difference—same with the stars.

The thing is, now from being this tight little universe, this poky little universe, we now suddenly have a huge universe and the earth is just a speck. The earth has no privileged position. It's not at the center and before long, people made the universe infinite and the sun was not at the center either. Suddenly, our home is not important. At the same time, all of these signs and things from the heavens are being given explanations—what we scientists call *natural* explanations. They were being shown not to be miraculous, but to be the effects of what we scientists call *laws*. These are not legal or moral laws like "Don't kill." They are scientific laws—unbroken regularities, like Newton's laws. They tell us about the way that the world works—the way in some sense that the world *must* work. Bodies fall to the ground. They must fall to the ground. Of course, not everything does fall to the ground. A boomerang doesn't. So then you look for reasons that make it an exception.

Science does not allow exceptions without reasons. That's why miracles are out. Miracles are exceptions because God makes them that way. And that is just not a scientific way of doing things—and a good thing, too. On the one hand, if you demand reasons, then usually you will find them. On the other hand, if you don't demand reasons, then soon enough, you are going to find fault and get into prejudice. Think of medicine in the nineteenth century. In cholera epidemics in New York, inevitably it was the poorest people who suffered most. So the rich Protestants all said, oh well, God is punishing these Irish for being Catholic. Now we know that it had nothing to do with being Irish or being Catholic or whatever. The poor people were drinking water that was contaminated by the toilets. The rich people were not. It was as simple as that. Or rather, it wasn't simple to find out, but it was there to be found out—a scientific answer.

But there's more than just that. I don't think of myself as a particularly modest man, but there's something modest about science and not about religion. Religions tell you, believe this or believe that, or go to hell. Hate homosexuals or you're doomed. Kill the infidel or you're doomed. Make women stay home or cover themselves in black robes or you're doomed. Like I said, I am a Popperian. Even the best science might be wrong. For three hundred years, nearly all people thought that Newton had got it right—absolutely, completely right, for all time. Then along came Einstein and oops, maybe the Newtonians weren't so right after all. And the same could happen to Einstein or the double helix or plate tectonics or whatever. You can always be sure that you are wrong. You can never be absolutely 100 percent sure that you are right. That's the modesty of science and that's the reason why science and religion are not only different but bound to be opposed. Science must be open to check—it must be "falsifiable." Religion is never open to check—it is "unfalsifiable." That is why science is for winners. Religion is for losers. But, more than that, science is for winners because it is more honest than religion—always was, always will be.

Matthews: I just have to say—

Rudge: A couple of points—

Wallace: I can see your background! David Davies. You really are a fiery Welshman.

Davies: Actually, I'm Jewish. My great-grandfather was called Davidovitch or some such thing. It got changed at Ellis Island. But I'm glad that point's come up. I am not a practicing Jew. In respects, I am not even that keen on Israel. But I can see the need for it. My ancestors spent generations being despised and bullied and used and beaten and killed—and then came the Holocaust. What was the main force behind all of this? Religion, religion, religion! We were thought to kill babies for our ritual practices. We were accused of murdering Jesus. We were . . . well, it makes me sick. Martin Luther was the world's biggest anti-Semite before Hitler—all done in the name of religion. For me, science is something clean, something that washes away that kind of thinking. It's difficult, it's hard, it's often disappointing. But, it's the grown-up way of doing things. For me, science isn't just something. It's a morally good something.

Fentiman: Right. Well, I'm going to intervene here. Next up is Hal Wallace. But I can see that Martin is bursting to say something. So, just as long as they are points of correction or amplification—

Rudge: Absolutely, because in many respects, I agree with Dave. I do think that science has swept away miracles. But I do want to say first that that stuff about the flat earth is simply nonsense. It's a fiction from the nineteenth century. Nobody in the Middle Ages thought the earth was flat. Everybody knew about the eclipse and the earth's circular shadow on the moon. Everybody knew that objects disappear over the horizon. The whole point of the Aristotelian world system is that the earth is a sphere, right at the center of another sphere that carries the stars. In fact, Aristotle's theory was known as the two-sphere theory.

Second, the medievals did not have a poky little universe. Before the Middle Ages, the great scientists had been the Arabs. They had worked out that the radius of the outer sphere of the stars from the center, from the earth, is 98 million miles in our units. Not much poky there, although I do agree that the big thing that Copernicus did was to extend the size of the universe hugely and that was significant, if not for the literal word, for the effect it had on people's imaginations. Just like Darwin and time. Oh, and incidentally, if you are going to throw in a third point, the Catholic Church—then and now—has always been violently against astrology. It is not a fan of zodiac signs. Astrology implies that our fates are sealed at birth, and that denies free will. And that's a no-no for Christians.

Fentiman: Well, thanks for that. Now, over to Pastor Hal.

Wallace: Of course, I want to disagree completely with Professor Davies. So much so, that you might think we stand back-to-back and share the conviction that science and religion must always be in conflict. Except, I know that religion has been the winner! Of course, I'm not a flat-earther, and I don't think the earth is the center of the universe. Copernicus was obviously right about this. I'm still not that keen on the Catholic Church, although, like a lot of evangelicals, I do think that Pope John Paul II was a truly great man. When he spoke out against abortion and against homosexual marriage—I just can't call it gay marriage because it just seems to me to be so sad—I think he was doing the work of Jesus. I absolutely do not think that God was punishing the Irish for being Catholic, although I am equally certain that God will punish each and every one of us who does not turn to him and acknowledge him as Lord. Finally, I am certain that I am a much greater supporter of Israel than Professor Davies. I just know that it is God's work that is happening over there.

I'm pro science and so is my congregation. We have a large number of doctors—really good ones, who are pediatricians and gynecologists and surgeons and so forth—and we have vets and chiropractors and others in the health-care business. Also, we have a large number of computer people of one sort or another. And this is not to mention our teachers, including our science teachers. My doctorate was given to me by Central Southern Baptist

Seminary, but I myself have earned a master's in nuclear engineering. I was all set to work in that field until my final year, when I found Jesus and decided to devote my life to him. So, we, in my church, are very much in favor of science. Even though I disagree almost completely with Professor Davies, I am not a conflict person like him. I don't think that science and religion have to be at war. I am much more of a dialogue person, or at least that is where I would think I would fit. Science and religion are both good things and they have much to say to each other.

So where do I differ from the professor? Well, of course, there are things I would want to say about religion. I think he's completely wrong to say that religion—Christianity, in particular—is responsible for the evils done to people, especially the evils done to Jews. If anything, the big influence was Charles Darwin, of all people! He led straight to Hitler, through the transfer of the struggle for existence to social issues. Hitler's own testament, *Mein Kampf*, often sounds like something from the *Origin of Species*. Might is right, nations struggling against nations, and the Germans being superior to everyone else. If you think of the real monsters of the twentieth century—Hitler, Stalin, Mao—they were all contemptuous of religion and wanted to substitute their own faith of one sort or another. But leave all of this. Let's get back to our topic of science and religion. One major problem—you might think *the* major problem—I have with Professor Davies' views is that I think he is just plain wrong about a lot of the science. His beliefs about biology, for instance. But since these will come up later—next week, perhaps—let's leave these too.

Rudge: What we can't leave is your easy connection between Darwin and Hitler. Every genuine scholar knows that matters are far more complex than you have said. For a start, recent scholarship suggests that the main influence on Hitler was a kind of German romanticism of the late nineteenth century—what is often centered on the German people, or *Volk*—something incorporating some half-baked philosophical ideas from Wagner and his operas, that was a real font of ideas for the Nazis. For a second, all of that stuff about societies struggling owes little to the writings of Charles Darwin and much more to others, notably the mid-Victorian man of letters Herbert Spencer. I know that ideas of this kind are often known collectively as "social Darwinism," but they are not very Darwinian at all. And, for a third, a lot of people who jumped from biology to society were anything but bloodthirsty fiends like Hitler and his crew. For instance, the Russian anarchist Prince Peter Kropotkin thought that humans and other animals have an innate tendency to help each other—mutual aid—and, hence, there is no need for societal rules and constraints. Crazy, you might say, but done in the name of biology.

As far as Hitler is concerned, no one thinks he did any systematic reading or thinking, and he just pulled phrases and ideas out of the air as it suited him. The autobiographical parts of *Mein Kampf* are mainly fiction, so why not the rest?

Fentiman: Okay, let's get back on track. Pastor Hal, you were telling us where you stand on the science-religion relationship.

Wallace: In a way, the really big issue is the whole question of *naturalism*. Phillip Johnson, the retired law professor and author of *Darwin on Trial*, has been my guide here. He distinguishes between what he calls "Methodological Naturalism" and what he calls "Metaphysical Naturalism." Methodological Naturalism is the assumption that the world—the physical world, that is—works according to unbroken law. There are no *super*natural forces at work. So if something strange happens, say a liquid changes from red to blue before your eyes, you know that it must be because of a chemical reaction in the liquid or—less likely perhaps—someone is manipulating the lighting in the room, or something like that—smoke and mirrors, as you might say. It is the working principle of scientists, or at least most scientists. In a dramatic sort of way of putting things, it is the methodological assumption of atheism. God or gods are not playing tricks. No miracles.

So much for Methodological Naturalism, now what about Metaphysical Naturalism? Metaphysical Naturalism is the belief that there really is no God or gods. There is nothing beyond what you see. Of course, in this day and age what you see can come through very complex and sophisticated machines. You don't look at a peach, say, the Georgia fruit, and see a DNA molecule. But the molecule is there, in some sense to be found through the senses. The Metaphysical Naturalist—someone like the biologist Richard Dawkins or the philosopher Daniel Dennett, and I suspect Professor Davies over there—says that that is all. The Metaphysical Naturalist really is an atheist.

Johnson refers to himself as a Theistic Realist. He means that he believes that God really truly does exist. As you can imagine, I am pretty happy with that myself. The point here, though, is the relationship between Methodological Naturalism and Metaphysical Naturalism. You might think, well okay, I'm going to be a Methodological Naturalist, but that says nothing about Metaphysical Naturalism. I'm going to keep God out of science, but I still worship and adore God and believe in his existence. A kind of independence view, if you like. But, Johnson argues, the point is that things don't really work that way. Once you get into Methodological Naturalism, then Metaphysical Naturalism is a step—a too-easy step—next. Someone like Professor Davies says, keep God out of biology, and the next thing he is saying is that there is no God. What he does from Monday through Saturday starts seeping into what he believes on Sunday. Although, I suppose that if he were a practicing Jew, it would be the Sabbath instead. But you know what I mean.

I'm not sure I want to go all of the way with Johnson. I think he's absolutely right that Methodological Naturalism generally leads straight to Metaphysical Naturalism and that's a bad thing. The question for me is how we should define or understand Methodological Naturalism to avoid this problem. I want to do science! I agree with Johnson that there are certain events in nature that simply do not come under the normal laws of nature. I guess we'll be talking more about these in the next program. But, do we want to say that these events are also part of science or do we want to say that science stops there and something else takes over? That something else, obviously, being God and religion. Some people, perhaps this includes the well-known Christian philosopher Alvin Plantinga, want to say that science should be expanded to include these events and their causes. The events are part of nature even if the causes are supernatural. He says that since everything is due to God all of the time—remember the hymn, "He's got the whole wide world in his hands"—we should not distinguish between events caused by law and events not caused by law. He calls this "Augustinian science."

Some people, and I feel happier about this, want to say that science should not include miracles, but that science should recognize more openly its limitations. Most Methodological Naturalists would not only deny the miracles of Jesus, but would deny the very possibility of the miracles of Jesus. If Jesus was born of a virgin, they would say that there must have been a natural cause, although probably they would say that because there couldn't be a natural cause—if the ovum started spontaneously to divide, since women carry only the X chromosome, the female chromosome, the offspring would have to be 2X, which is female—Jesus couldn't have been born of a virgin. I want to stop that move by saying that Methodological Naturalism has its limits. The Reverend William Whewell, a nineteenth-century Anglican clergyman, used to say about the origins of organisms that "when we inquire whence they came into this our world, geology is silent. The mystery of creation is not within the range of her legitimate territory; she says nothing, but she points upwards." I agree with that exactly.

I would say that I am all in favor of Methodological Naturalism, but that it's not as powerful as most people think.

Rudge: You mention Plantinga. He thinks that sometimes science just can't explain things, and he says that's okay. People like Dave and me want to say that that is just the attitude that you should not have. You give up too soon, when there might be a solution. We say that his attitude leads to premature "science stoppers." Can't find an answer? Must be a miracle! We say: nonsense! Keep working and the answer will come. Even if it doesn't, the answer's there.

Wallace: But that just proves my point. You people define out the very possibility of miracles. Science stoppers may be a problem for science, but whoever said that Almighty God had to go along with the rules of the National Science Foundation?

Fentiman: Can I just get a word in here? To move things along, Reverend Hal, where do you locate yourself on the spectrum? You said something earlier about not being a conflict person—

Wallace: I am getting to that. I think that science is science and religion is religion. They are not the same. But, they do certainly complement each other. So, there's dialogue. It's the old metaphor of two books, the Book of the Bible and the Book of Nature. Another hymn, this one incidentally by a great Anglican, John Keble:

> There is a book, who runs may read,
> Which heavenly truth imparts,
> And all the lore its scholars need,
> Pure eyes and Christian hearts.
>
> The works of God above, below,
> Within us and around,
> Are pages in that book, to shew
> How God himself is found.

Fentiman: I am a little sorry that you didn't sing it for us! Perhaps next time we can try for a barbershop quartet! Now, let's turn to our other minister, the Reverend Emily Matthews. I take it—goodness, what do I call you? I can hardly call you Father Matthews. Mother Matthews?

Matthews: Why don't we just settle for Emily. I am glad to get my turn, because obviously I'm going to disagree with David Davies, but I'm going to disagree with Hal, as well. I'm much more into integration. Let me say right from the beginning that I have been very much influenced by Alfred North Whitehead and I think that his "Process Philosophy" is the right way forward, theologically. I realize that this philosophy can be horrendously difficult to follow, particularly with all of the new words that its devotees have coined, but I understand it to mean that all things are moving, are in a process of development or becoming. This applies to God himself, as well as to his creation. Obviously, this means that I reject some of the traditional Christian views of God. Saint Augustine said that God is all-powerful and all-knowing, and that in some sense, he is eternal. He stands outside time and is unchanging. I think this is completely misguided, and reflects Augustine's enthusiasm for Greek philosophy—particularly the philosophy of Plato, who lived four hundred years before the Common Era—rather than the message of Jesus of Nazarus, who is the Christ, who is the son of God. For Whiteheadians, God is not all-powerful in the traditional way. Rather, God is in the process along with humans. He can help, he can influence, but he cannot determine. In a sense, as the Lutheran theologian Philip Hefner has said, we humans are cocreators, along with God.

I want to stress that I don't think I am just modifying Christianity to suit my own ends. I think that Augustine was wrong. He was wrong in a lot of things, but especially here. He ignores entirely the notion of "kenosis." This is the idea that God in some sense empties himself, he voluntarily—especially in the person of Jesus—makes himself limited. Philippians 2:7 says explicitly that "Jesus made himself nothing" or "he emptied himself." I think that this is the only way we can explain the agony on the cross. God the Father did not have to die. God the Father is eternal and cannot suffer. But God came down to earth, made himself human—made himself God the Son—and put himself in the order of time, and suffered and died on the cross. If that is not making oneself nothing, I don't know what is.

For me, this kind of thinking is the only way I can make sense of the terrible events of the last century. How does one explain Auschwitz? How does one explain the death of those children in the camps? How could a good god have let Anne Frank die of typhus at Bergen-Belsen? Only if you are prepared to say that God was with them at Auschwitz. God, too, died of typhus at Bergen-Belsen. The great Jew Elie Wiesel says that God died at Auschwitz. He did, but he has risen again and continues to share our suffering and work with us to bring things to completion. For me, it was not just chance that

Anne Frank wrote her diary and that it was found and has been one of the truly inspirational works of our time.

Rudge: This is all very well, and I agree that it is a noble vision—although I am not sure that Anne's death is quite balanced by her book and its influence—but, in any case, I don't see what it has to do with science and religion. In what sense are you an integrationist? Hitler killing Jews was not science. So how does making this compatible with your theology make for an integration of science and religion?

Davies: I would say that Hitler killing Jews was science. Bad science, namely assuming that Jews are different from other human beings, especially Germans.

Matthews: Actually, I'll want to pick up on that point in one of the later programs. But for now, let me speak to Martin's complaint. It's true, I haven't yet got to integrating science and religion. It's at this point I want to bring in feminism.

Davies: Here we go. I knew we'd get to that sooner or later, probably sooner.

Matthews: And I knew that sooner or later, probably sooner, we'd get your prejudices out on the table! Although, I expect you're not the only one. You have twenty-three pastors in your church, Reverend Hal. How many of them are women?

Wallace: As it happens, we have not yet felt the call to ordain women. Although, we do have many theologically qualified women active in our youth groups. We are very proud of the fact that we have more women on our staff with doctorates than the men.

Matthews: Proud? That's not the word that comes first to my mind. But let's leave that for now. I suspect these issues will arise again before we are done. Let's get back to my position. The historian Carolyn Merchant has shown that the so-called Scientific Revolution—the Copernicus-to-Newton stuff—was much more than a simple scientific change.

Rudge: I wouldn't exactly say the change was simple!

Matthews: Fair enough—was much more than a straight scientific change, however significant. The really big thing was the change of metaphors. Up through the Middle Ages, the dominant metaphor had been the world as an organism. People thought that the globe on which we live was living—it had life, it had a soul, the *anima mundi*. This was a very old idea that goes back in fact to Plato—so I'm not against Greek philosophy generally—that says that the earth is an organism. It's all in the dialogue the *Timaeus*, which was by far the most popular dialogue in the medieval period. It's not a silly idea by any means. The earth goes through phases—winter, spring, summer, and so forth—just like we get up and live the day and then sleep, and then start all over again. The earth is nourished by the sun. The springs of water well up and form rivers going down to the sea, and fertilize the land. You dig into the earth and you find its veins—think how we use just that word for things carrying blood and for seams of coal and metal—and so on and so forth. It was of course deeply feminine—the earth is always our Mother and never our Father—because we humans are nourished by the earth and we are born from it.

Then all of that was changed, and the organic metaphor was rejected. Thanks to Galileo and Newton and the others, especially Descartes, the earth was no longer an organism. It was a machine. It was a clock that worked according to fixed rules or laws, that was predictable, and that was in itself not living. It was created by God, that is true, but the earth itself was now lifeless. And, of course, this meant that you could do anything you wanted to it. If you were damming up a river, you had no need to pay respects to the earth's needs or welfare. If you were digging a mine, you did not have to worry about violating the earth's innards. It was simply there for our use and exploitation.

Of course, today we all see what a disastrous metaphor that has all proven to be. Global warming is just the start. Vast areas of the world have been reduced to barren rubble or desert, unfit for human habitation. Jungles have been cleared and destroyed. The seas are polluted beyond belief. It used to be that off the coast of Newfoundland, the fishermen could virtually walk on the sea, there was so much cod out there. Now the seas are closed. The fish are gone. The waters are silent. And it is the same in one place after another. DDT robbed us of the birds. Fertilizers rendered the soils barren and made dust bowls of the Midwest. Mining has stripped the lands and the coal, and oil has polluted the atmosphere. We need to respect and love our Mother. Thank God—and I mean thank God—some people now recognize this. Speaking now as an ardent ecofeminist—a woman who respects and loves her Mother—my sisters and I have now embraced the *Gaia hypothesis*. This is the brainchild of the great English scientist James Lovelock, a Fellow of the Royal Society, who argues that the world is an organism and we should respect and love it.

Like everyone else, I, too, have brought my passage to quote. This is from Lovelock's *Gaia—A New Look at Life on Earth*. He writes: "The entire range of living matter on Earth, from whales to viruses and from oaks to algae, could be regarded as constituting a single living entity capable of maintaining the Earth's atmosphere to suit its overall needs and endowed with faculties and powers far beyond those of its constituent parts." He goes on to say that Gaia can be defined as "a complex entity involving the Earth's biosphere, atmosphere, oceans, and soil, the totality constituting a feedback of cybernetic systems which seeks an optimal physical and chemical environment for life on this planet." The other person who is developing this is Lynn Margulis.

Davies: Oh, the ex-wife of Carl Sagan.

Matthews: I'm not going to rise to that. In this discussion, Lynn Margulis is no one's wife or ex-wife. She is one of the most brilliant scientists of our age. I am sure that this will come up at a later point.

Fentiman: So, pull it together now.

Matthews: My position simply is that we live in a developing world, and that God is part of it. He lives throughout all reality.

Wallace: Sorry, I've got to get in here. That is total heresy. God is not the world. The world is his creation. He is separate from this. To argue otherwise is to be a pantheist, like the Dutch philosopher Spinoza.

Matthews: I'm not a pantheist, and indeed I'm not sure that Spinoza was a pantheist. Technically, I am what is known as a "panentheist." The American philosopher Charles Hartshorne wrote about this. I think that God is everywhere, but that he is sustaining the Creation. He is with of it, but not part of it. The world is an organism. It exists independently, just as we humans do. God strives to bring it to fulfillment, but the world and we humans must work with God to realize the ends. In this sense, science and religion work together. If you think about it, I couldn't be anything but an integrationist. I'm a holist, I'm an antireductionist. I'm an emergentist. The whole is bigger than the parts!

Davies: The trouble with people like you is that you just keep chopping and changing things to suit your needs. Now, God is eternal. Now, he isn't! Now, God is all-powerful. Now, he isn't! I thought religion was supposed to be about the eternal verities. How can it keep changing as you suggest?

Matthews: Giving you the courtesy of thinking that this is a serious question and not just a rhetorical attempt at refutation, I am going to give you the answer I once heard Langdon Gilkey give to a professor of anatomy who was ribbing him on just this point. He asked the anatomist: "Well, tell me, how is your Department of Bleeding going these days?" The anatomist pulled up in indignation and replied: "We haven't had a Department of Bleeding since the eighteenth century." To which Gilkey responded: "Then why do you keep claiming that religion and theology must always stand still? Science and medicine change over the years as we get better understandings. Why should it be any different for knowledge of the Divine? In both cases, as new ideas and new techniques come in, we try to give a better account than previously." And that is precisely how I think I can move beyond Augustine and Aquinas and on to Whitehead. He was not more brilliant than the earlier thinkers, but he did have many hundreds of years of thought and discovery to build upon.

Fentiman: Well, after that, anything is going to be a bit of an anticlimax. Sorry about that, Martin Rudge. But, it's your turn now.

Rudge: Don't feel sorry for me. Like the marriage at Cana, the best is left until last.

Davies: Explain.

Rudge: The Gospel according to John tells us that the first miracle Christ performed was at Cana. He was at a wedding and the wine ran out. At the urging of his mother, Jesus had the pitchers filled with water and turned them into wine. The point was that they were better than the wine that had gone before. So when you say that something is like the marriage at Cana, you mean the best is left until last. But, when you spell it out like that, it's not that funny. So I won't try any more jokes like that.

Davies: Thank goodness. In any case, I thought you were an atheist.

Rudge: Actually, I think of myself as more of an agnostic—I really don't know the answers—although I prefer to call myself a "skeptic." Agnostics so often strike me as people who don't really care about the issues. I care desperately; it is just that I don't have the answers. And by the way, although I realize that my Cana joke may not have been that funny—at least, when it was spelled out for people like David Davies—I am not going to apologize for making biblical references. It may not be very politically correct, it may violate the separation of church and state—and I, of all people, think that that's important—but, the Bible is part of our cultural heritage. Dead white males and all.

So to be perfectly candid, and no doubt to be insulting and certainly to be guilty of the same sin all over again, I think that anyone who does not know the basic biblical references is a bit of a philistine. We all live in a society that, for all its faults, gives us freedom and the time and health to follow our interests. In return, we should understand and appreciate the beliefs and references of those who made it possible. And no one can know anything about the history of the United States without knowing about the key significance of Christianity. It was this that sustained our predecessors through the awful and hard and boring and dangerous mission of making America the land that it is. So, we all have an obligation to know about the beliefs of our founders, and I don't just mean those that signed the Declaration of Independence. Oh, and by the way, Dave, I'm not ramming my beliefs down your throat. Philistines come from your part of the Bible.

Fentiman: Right, folks. Great stuff, but let's calm down and keep to the topic. Martin, where do you stand on the science and religion issue?

Rudge: It seems almost too pat to say this, but I'm drawn strongly to the one slot still left empty, the independence option. I agree 100 percent with David Davies that science and religion can clash. Here, we are echoing Richard Dawkins, who has stressed this point strongly. I think one of the worst books I have ever read is Richard Dawkins's best seller *The God Delusion*. The level of discourse would make a bright sophomore cringe. But agreeing with Dawkins is not always a silly thing to do. In fact, when it comes to Christianity—son of God, died on the cross and then resurrected, and so forth—I probably have no more belief that Dawkins or Davies. Belief or not, there's a clash. The fact is that you cannot believe in the modern theory of evolution and be a Young Earth Creationist—that is, someone who thinks the earth is only six thousand years old, that the total creation took six literal days, and so forth. That's just not on. My favorite example of a clash comes from the Mormons—the Church of Jesus Christ of Latter-day Saints out in Utah—and their beliefs about North American Native people. They think that the "Red Indians" are the descendants of some of the lost tribes of Israel—Ephraim and Mana, to be precise—and that those left are the dark-skinned and degenerate "Lamanites." In the light of modern science, which shows absolutely no biological connection between Jews and American natives, but serious links to peoples who might have entered North America from Asia via Alaska, you simply cannot take the Mormon claims seriously. You either go with modern anthropology or with Joseph Smith and Brigham Young, but not both.

So, conflict is right on the table. That is established. The question is whether that is the end of the story, as Dave here argues. I don't think it is. Remember that I don't want to defend Christianity—let's stay with that—as something true and real. I have already said I myself don't think it is. The point is whether someone can be a Christian—someone who does think that Christianity is something true and real—and accept modern science all of the way.

Obviously, it all comes down to what you mean by "Christian." Given what we've heard already in this discussion, that point is almost a truism. If you are like Hal, then whatever he may say, I think you are plunged into the conflict almost immediately. I've looked at his church website and I know that as part of their faith commitment, they single out evolution explicitly as false. But leave him and the Southern Baptists, and go the other way. If you have a kind of watered-down Unitarianism, then you can obviously accept all of science. If you say that you don't know if God exists and if he does, it is all subjective—God is what God means to you—and Jesus was just a good guy and the miracles did not occur or if they did they were natural—a passing comet parted the waters of the Red Sea—then you can have all of the science that you want. It seems to me that this is the position of people like the biologist Ursula Goodenough. They want to have a kind of spirituality without the spirit, or at least without the Spirit with a capital S. You can hug trees and can get into the Gaia hypothesis, and have physics, chemistry, and the lot. It helps if you talk in deep tones about things like emergence and holism.

Davies: What is this "wholism" that you keep talking about?

Rudge: Sorry, it is one of those words like "hegemony" and "hermeneutics" that intellectuals like to throw into conversations today, without much knowing what they mean. It was a term invented by the South African statesman Jan Smuts, who fancied himself as a bit of a philosopher. It means that you cannot understand something by just looking at the parts. You must look at the whole. In other words, it is the opposite of "reductionism." Incidentally, to confuse things even more, it is spelled without a beginning w, as *holism*.

Back to the point I want to make. I just don't think that watered-down stuff about feelings and hugging trees and wearing herbal sandals is Christianity. It is thin to the point of nonbeing. I may be a nonbeliever, but I'm a pretty conservative nonbeliever. I think the Christian, at a minimum, has to accept God, Jesus as the son of God, death on the cross and subsequent Resurrection—whatever that might mean—forgiveness of our sins, and the possibility of eternal salvation. That's my quarrel with Stephen Jay Gould. In his little book, *Rocks of Ages*, he talks of science and religion being two *Magisteria*—ways of looking at things that cannot by nature clash. But then, he waters down religion to vague ethical sentiments. People who want to believe in things like the Resurrection are apparently silly. As far as I am concerned, that is the whole point. The Resurrection may be false, but I want a space for people to believe in it as well as science. False, but not necessarily silly. So my question is whether you can have a genuine traditional Christianity—remember I am staying with this religion, and by traditional I mean the Christianity of Augustine and Aquinas, Luther and Calvin too, if you like—and hold on to science?

I think you can. Traditional Christianity has never insisted on an absolutely literal reading of the Bible. It has always been understood that interpretation is needed, and sometimes is absolutely necessary. Saint Augustine was clear on this. Even in his day, four hundred years after Jesus, there was still debate among Christians as to the attitude they should take toward the Old Testament. This collected the sacred writings of the Jews, but the Jews were the people who had urged the crucifixion of Jesus and rejected him as the messiah. Some groups or sects, including one known as the Manicheans, thought strongly that, given history, Christians had no business accepting the Old Testament, and they made almost a profession of finding flaws in the writings, such as the two inconsistent accounts of the Creation, as given in Genesis 1 and 2.

As a young man, Augustine was attracted to the Manicheans and, like them, he rejected the Old Testament as false. When he converted to Christianity, as part and parcel of the deal (as one might say), Augustine had to reverse himself on the veracity of the Old Testament. There were good reasons to do this. Most importantly, the Old Testament makes sense of the New

Testament: Why did Jesus have to die on the cross? To wash away our sins. Why are we, the creation of a good god, sinful? Because Adam ate the apple. But, having to this point spent his time poking fun at literal interpretations of the Bible, Augustine, always warned his fellow believers of the problems with literal interpretations. It is true that the great reformers, Luther and Calvin, were much keener on reading the Bible as it is, but they, too, recognized the need to interpret. Calvin spoke of God accommodating his language to the common people. There was no point in God using the terms of sophisticated science because no one would have been able to understand it.

Davies: Yes, but even if you allow that Christians have the freedom to interpret the Bible to such an extent as to allow modern science—and I should say that I'm still not absolutely convinced on this—you still have the miracles. Physiology tells us that when somebody dies, they die. What about Lazarus and Jesus? It seems to me that this is just as much of a conflict with science as the Mormons and their crazy ideas about the Navajo.

Rudge: Actually, although there were Navajo in the southern part of what we now call Utah, the Omaha were the first Indians that the Mormons encountered on their trek west. Then, hardly surprisingly, it was the Utes. But, get back to miracles like the Resurrection. That seems to me to be a really important question. I think you can go two ways on this, and it all rather depends on your interpretation of "miracle." So far, we have been talking as if miracles always mean breaking laws. But that's not quite accurate. If you are in the tradition of Augustine, then you downplay the breaking of natural laws aspect and concentrate on meaning. A miracle is less something that is a violation of physics, say, and more something that means something to people at the time or later. If someone is cured of an illness after prayer, it's irrelevant whether or not God broke the causal chain or whether there was a real medical reason for the spontaneous cure. The point is that it means something; namely, that the person was cured after prayer.

The same is true of the Resurrection of Jesus. It is irrelevant whether physically he rose from the dead. Don't get into the business of thinking that perhaps he was in a catatonic trance when he was taken down from the cross or whether his stinking body was or was not in the tomb on that Sunday morning. The point is that the disciples, who were downcast, were suddenly filled with hope and love, to an extent that they went forth to spread the gospel. And if someone says that this is a well-known psychological phenomenon, then so much the better. It simply doesn't count against the meaning of the Resurrection.

The other way, usually associated more with Aquinas, is to accept that miracles do indeed involve the breaking of law. Note that this does not always cause problems. We've already mentioned transubstantiation, the turning of the bread and the wine into the body and blood of Christ during the Catholic Mass. This doesn't break laws as we can see them, because no one ever says that the breadlike and winelike substances disappear. More generally, however, what you have to do is invoke the distinction between the Order of Nature and the Order of Grace. Normally God works through unbroken law. But when it came to human salvation, where he himself had to come down to earth and become human, it was necessary to have lawbreaking miracles. This is the Order of Grace. Don't look in your funeral parlors for the spontaneous reappearance of life, because it just won't happen. But, it was necessary that it happen this once—more if you take other miracles into account—a couple of thousand years ago in the land of Palestine. Some Christians, of course, rather muddy all of this by assuming that lawbreaking miracles are still going on, but I can only defend the position so far. Although, note how reluctant the Catholic Church is to accept miracles and visions and so forth.

Fentiman: I'm still not sure, Martin, that you've told us why you're an independence person. Let's agree that you've made a space for science and religion, but you haven't really told us why you think they must be independent.

Rudge: Fair enough. My position, and we've already mentioned people like Karl Barth on this—in fact, I think that Gould was right in principle if not in execution—is that science and religion speak about different things. This is where I am in square opposition to Richard Dawkins in *The God Delusion*. Science and religion can be in opposition—Noah's Flood is simply inconsistent with modern geology—but, essentially, Dawkins is just wrong when he claims that religion is an empirical matter. God—the God of the Christians and the Jews and the Muslims—is not something unseen like a dinosaur or an electron. If he does exist, then he is wholly other. For a start, we humans are contingent. Hard as it may be to imagine, David Davies might never have existed—just not there at all. God, however, is a necessary being. He stands outside time and is, thus, eternal. Just as $2 + 2 = 4$ is necessary—it never became true and it will never become false—so God exists necessarily. He is not just very old, but beyond age. You might think that this does not make sense, but it is part of Christian theology. This also defines our relationship to God. Because he is necessary and we are not, we must have been created by an act of free will. We do not follow as a matter of course, like Pythagoras's theorem follows as a matter of course from Euclid's axioms. God did not have to make us, but he did—out of love, argues the believer.

Hence, the whole business of putting God claims on the same level as material claims is what the philosophers call a "category mistake." It is like asking if Tuesday is tired. Tuesday is simply not the sort of thing that can be tired or not tired. Likewise, God claims have to be different from claims about the physical world, the object of science. Langdon Gilkey—Emily has mentioned him already—who was a theologian for many years at the University of Chicago School of Divinity and was probably the most eminent spokesman for the neo-orthodox position in America in the second half of the twentieth century, used to speak of "how questions" and "why questions." Think of somebody asking you what you did yesterday afternoon. How did you fill your time? You could say: "Well, I went on a picnic with my family, and we took the Chevy, and we ate ham sandwiches and drank Coke. Then we paddled in the stream, and then we went back home and got caught in a traffic jam." Or you could talk about why you went on the picnic: "My kids and I put on a picnic in order to celebrate my wife's birthday. We wanted to show her how much we love her and how appreciative we are of what she does for us." Two different answers—both perfectly good, but about different things. They not only don't clash, they couldn't clash. The same is true of talk about the physical world and talk about God. Chalk and cheese, as we've said.

Davies: I'm not sure about the not-clashing bit. If you told me that you'd gone to a sewage farm then, unless it turns out that your wife is a sanitation engineer, I wouldn't accept that you were celebrating her birthday. Or perhaps you philosophers are even odder than I already think!

Wallace: I'm still not convinced that you've really gone much beyond Gould. You know where I stand—at least, I guess you do, since you've been looking at my church's website. I want to say that God is creator of heaven and earth. I want to say that humans are made in God's image. I want to say that God laid down laws of conduct and that we humans disobeyed him and that we are sinful. I want to say—and you yourself have agreed that as a Christian I should say this—that because of Jesus's sacrifice, there is a place for us in heaven and hope of eternal salvation. What about all of this? How can you make room for all of this, given your understanding of science? More than this, and here I may be agreeing with Professor Davies of all people, I am uncomfortable with making God so wholly other. Saint Thomas Aquinas said we can know God analogically. He is not our physical father, but we know him as a father—creating, loving, caring, and so forth. Tell me how this fits into your theology.

Rudge: Well, it's not my theology, but what I take to be conventional Christian theology. I would say—

Fentiman: Sorry, I am going to have to cut off the discussion here. Let's hold this thinking, if you don't mind. I am afraid that today we are out of time. From, starting at my far left, David Davies, Martin Rudge, Emily Matthews, and the Reverend Hal Wallace, this is the host of *Eternal Questions*, Redvers Fentiman, wishing you good night, and inviting you to join us next week, when the same panel will be turning their attentions to the evolution and creation controversy.

~

CHAPTER TWO

Program Two: Origins

Fentiman: Hello. My name's Redvers Fentiman, and I am the host of *Eternal Questions*. We come tonight to the second program of five devoted to the topic of "Science and Religion: Who Is Winning?" My panel of experts is back from last week, and having warmed up on a general discussion of positions one might take on the science-religion relationship, now we start to get a bit more specific.

Rudge: Just before we start, can you settle an argument for us? Waiting to go on, we were speculating about your somewhat unusual first name. Where does it come from and why do you pronounce it "Reevers," without a *d*?

Fentiman: Good question, and I don't have all of the answers. I can tell you where it comes from. My great-great-grandfather was the soldier-servant, the batman, of a British general, Redvers Buller V. C. Poor Buller, I am afraid, was a brave soldier but a lousy leader, and led his troops to colossal failures in the Boer War, just at the end of the Victorian era a hundred years ago. Nevertheless, we Fentimans have been loyal to his memory, and his name got passed on down to us. My grandfather emigrated to the States between the wars, and the name came with him. Why Reevers rather than Redvers I just don't know. Not quite as bad as Featherstonehaugh, which is pronounced Fanshaw!

In fact, Redvers is not a made-up name, but it is rare. The British author C. P. Snow wrote a book with a character who has the first name Redvers, although I don't know how he pronounced it! The book is called *The Masters,* and it is about the election of the head of a Cambridge college. Heads are often called "Master." It ties into today's topic, because although the college in the novel is fictional, it was really meant to be Christ's College, which was Snow's old college. And it was Christ's College that Charles Darwin attended back in the first part of the nineteenth century. Charles Darwin is the main figure in our discussion today, but, whether he is a hero or a villain, I guess, remains to be seen.

What we can say and all agree on straight off—probably the last moment that we will agree in the next hour—is that there is a huge row in American society today over the theory of evolution. And rightly or wrongly, Darwin is the person most identified with this row. Since the critics are almost all Christians, we have the science and religion controversy in spades. This is the topic of our series, and evolution—for and against—is the topic of tonight's segment. Again, I think it makes most sense to start off with Professor David Davies, our biologist from the Massachusetts State Institute of Science. Why don't you tell us a bit about evolutionary thinking and why you support it?

Davies: First the basic picture, and then let's focus in on the theory. The world is filled with organisms: plants, animals, and literally billions of microorganisms—living things too small for the human eye. Last week, Emily worried about the death and destruction of modern life, and she surely has a point. But there are still huge numbers of organisms, divided into many, many different kinds. The evolutionist claims that these did not just appear separately. Today's organisms, and all of those revealed in the fossil record, have a common ancestry going back huge numbers of years. There is one great big tree of life connecting everything.

We can put some absolute dates on these sorts of things. The universe itself is about fifteen or so billion years old. The earth and the moon are about four and a half billion years old. Initially, the earth was boiling hot, and then it started to cool. As soon as it was possible for life to appear, it did appear. This was about 3¾ billion years ago. The last universal common ancestor of life on earth was about 3½ billion years ago. The really exciting stuff started about half a billion years ago, or more accurately 542 million years ago. This was the so-called Cambrian explosion, which lasted thirty or more million years when we got a huge number of new complex forms appearing all over the world. Some of the most famous fossils are those little marine invertebrates, the trilobites. Vertebrates started around then. Those are organisms with backbones. From then on, it was one evolutionary success or advance after another. The fish appeared. They left the water, and we got first am-

phibians and then reptiles. Some of the reptiles developed into the most famous fossils of all, the dinosaurs. Others went different routes and about two
hundred million years ago became mammals, little creatures with warm,
hairy bodies that suckle their young. But it wasn't until the dinosaurs were
wiped out, probably by a comet hitting the earth about sixty-five million
years ago, that the mammals really started to flourish.

Except of course, as we all now know, the dinos were not completely
wiped out because some of their successors are still around. Today we call
them "birds." The human line appears somewhere around five million years
ago, just after we had broken off from the chimpanzees, our closest relatives.
Then they were just tiny little creatures with small brains. It wasn't until less
than a million years ago that humans, as we understand them today, actually
appeared. Almost certainly in Africa, and then they traveled. Although, re-
markable as it might seem given the differences today between humans, it
was not much more than a hundred thousand years that the real trips and
separations began.

Matthews: I thought that the differences between humans weren't that great.

Davies: Actually, you are right. The differences are not that great as differences go, although they are great enough to separate races. But I guess that
is something we'll come to later. For now, let us push on and talk about
causes. I don't think that anyone would ever say that there is only one cause
of evolutionary change, but every professional biologist would agree that one
cause stands head and shoulders above all others. This is natural selection.
This is the idea that was expressed by the great English naturalist Charles
Darwin in his *Origin of Species*, back in the middle of the nineteenth century.
I've brought the book along—at least, I've brought a reprint of the book
along—because I thought it would be worth quoting from the *Origin* itself.
There are two parts to Darwin's argument. First, he argues to something
called the "struggle for existence."

> A struggle for existence inevitably follows from the high rate at which all or-
> ganic beings tend to increase. Every being, which during its natural lifetime
> produces several eggs or seeds, must suffer destruction during some period of its
> life, and during some season or occasional year, otherwise, on the principle of
> geometrical increase, its numbers would quickly become so inordinately great
> that no country could support the product. Hence, as more individuals are pro-
> duced than can possibly survive, there must in every case be a struggle for ex-
> istence, either one individual with another of the same species, or with the

individuals of distinct species, or with the physical conditions of life. It is the doctrine of Malthus applied with manifold force to the whole animal and vegetable kingdoms; for in this case, there can be no artificial increase of food, and no prudential restraint from marriage. Although some species may be now increasing, more or less rapidly, in numbers, all cannot do so, for the world would not hold them.

Then in the next chapter, he goes on to natural selection.

Let it be borne in mind in what an endless number of strange peculiarities our domestic productions and, in a lesser degree, those under nature vary, and how strong the hereditary tendency is. Under domestication, it may be truly said that the whole organisation becomes in some degree plastic. Let it be borne in mind how infinitely complex and close-fitting are the mutual relations of all organic beings to each other and to their physical conditions of life. Can it, then, be thought improbable, seeing that variations useful to man have undoubtedly occurred, that other variations useful in some way to each being in the great and complex battle of life, should sometimes occur in the course of thousands of generations? If such do occur, can we doubt (remembering that many more individuals are born than can possibly survive) that individuals having any advantage, however slight, over others, would have the best chance of surviving and of procreating their kind? On the other hand, we may feel sure that any variation in the least degree injurious would be rigidly destroyed. This preservation of favourable variations and the rejection of injurious variations, I call Natural Selection.

Matthews: I thought he called it the "survival of the fittest."

Rudge: If I can get in here, "natural selection" was Darwin's original term. Obviously, he took it by analogy with the selection that breeders do to get bigger cows and shaggier sheep and better cabbages. Then, the man who also independently discovered natural selection, Alfred Russel Wallace, said that natural selection had too much of a God aura about it—artificial selection by humans and natural selection by God—and suggested that Darwin use the alternative term, the "survival of the fittest." This second term was due to Herbert Spencer. Darwin added the term to later editions of the *Origin*, so you can always tell when someone is using the first edition, it was published in 1859, because it doesn't use the term "survival of the fittest."

We should also add a bit about the Malthus stuff. The Reverend Thomas Robert Malthus was an Anglican clergyman at the end of the eighteenth century who was worried about what he thought were false philosophies of

progress—philosophies that suggested that humans were going to rise upwards and get better and better all the time. Especially after the French Revolution, he thought that this was dangerous pie in the sky. So, he pointed out
that food supplies would always be outstripped by population pressures and
that this would lead to consequent "struggles for existence." This is what
Darwin picks up and uses. His genius is to take a profoundly nonchanging
view of life and turn it into the basis of a life-picture that is always changing.
The reference to "prudential restraint" came from a later edition of Malthus's
work—one that Darwin read—where Malthus was trying to counter criticisms that surely God made it possible to avoid the struggles. Malthus, somewhat reluctantly, allowed that perhaps population numbers could be kept
down by "prudential restraint." The laugh is that the novelist Aldous Huxley, the grandson of Darwin's great supporter Thomas Henry Huxley, spoke
of Malthusian practices in his great satire Brave New World. By "Malthusian
drill," Huxley meant putting in your diaphragm! Malthus thought that that
sort of thing was disgusting and immoral. Malthus meant not marrying and
not doing forbidden things on the side, like consorting with prostitutes or
trying to prevent conception.

Fentiman: Fascinating though it may be to learn about Malthus and the con-
traception issue, let's get back to David Davies. What about evolution today?
I've heard that there's a lot of criticism of natural selection by people like
Stephen Jay Gould. Didn't he introduce another theory, punctuated equilibrium or some such thing?

Davies: One step at a time. Two steps actually. Most of the Origin of
Species is less about proving natural selection and more about applying it.
The great strength of the work is the way in which Darwin goes through
all of the various branches of biology—instinct and behavior, paleontol-
ogy, biogeography or the distributions of organism, systematics or classification, anatomy and morphology, embryology—showing how an evolutionary interpretation throws so much light on the branches or areas and
conversely how the branches support evolution through selection. For instance, Darwin asks why it is that the birds and reptiles (especially the giant tortoises) of the Galapagos archipelago, a group of islands on the equa-
tor in the Pacific, are similar from island to island and yet different. His
answer obviously is evolution through natural selection! Likewise Darwin
used his mechanism to explain why it is that the Galapagos inhabitants
look like those of South America, and the Canary Island inhabitants, in

the Atlantic, look like those of Africa. The inhabitants came from their respective closer continents and then evolved.

In a way, Darwin's argumentation was very much like that used in a court of law. We don't see the butler kill the lord, but we infer it from the clues—the bloodstains, the busted alibi, the motive. All of these point to the butler's guilt. Conversely, the butler's guilt explains the clues—same with evolution through natural selection.

Rudge: I might add that Darwin didn't just back into this method of reasoning by chance. His mentor was the historian and philosopher of science, William Whewell—incidentally, Hal, although the name is spelled "Weywell," it is pronounced "Hule" with silent w's. Another of those awkward English names to pronounce, I am afraid. Whewell was concerned with showing that you did not need direct evidence of actual rays to accept the wave theory of light. You do it all indirectly, through the clues, putting the cause at the connecting point of a fanlike net of reasoning. Whewell, who rather liked big terms, called this a "consilience of inductions." When he was challenged, after the *Origin*, Darwin always mentioned the wave theory.

Davies: Back to Darwin. The second point to be made about Darwin's theory is that he didn't have a proper theory of heredity. Unless you've got that, natural selection won't work. Think of it this way. Suppose my wife and I have four kids and Martin here and his wife have four as well. Suppose that, on average, half the kids get killed or don't reproduce. So in other words, the numbers stay the same. Four people have four active kids. Note, incidentally, that more important than survival and existence is reproduction. There's no point in having the physique of Tarzan if you're not interested in sex. Better to be a weedy little runt with a strong sex drive.

But I digress. In the next generation, there are four kids ready to reproduce to match the four parents. Suppose—notice that this is the way that science works, we keep supposing—we call it building "models"—suppose that three of the kids are Davies kids and only one is a Rudge kid. In other words, one Davies child did not make it, but three Rudge children fell into that category. It looks like natural selection at work. What Darwin says is that it is only natural selection if it is not a matter of chance that there are three Davies kids to one Rudge kid. The Davies kids have to have something—some feature, what we evolutionists call an "adaptation"—that makes the Davieses better reproducers. That makes the Davieses "fitter" in today's lingo.

Let's say the feature is—oh, I don't know—let's say it is a kind of camouflage. We all live on a dark surface—rich soil—and the Davieses are dark, but the Rudges are light, and the big problem is a predator. Say a giant bird like the Roc that carried off Sinbad. Sorry, these stories tend to sound like Hans Andersen. Or should I say Hans *Christian* Andersen?

Wallace: They sound like Hans Andersen because that's what they are. Fairy stories!

Fentiman: Well, that's what we're here to find out. Go on, David.

Davies: The point is that the Davieses have an adaptation, dark skin, which leads to their success in natural selection. The group is evolving toward Davies-like features. But only on one condition! The skin colors must be heritable. I mean, dark-skinned people have got to have dark-skinned kids and light skins, light kids. If the Davies kids were to have only light-skinned kids and the Rudge kid to have dark-skinned kids, selection would be all over the place. There certainly couldn't be any evolution. And this was Darwin's problem, the weakness in the *Origin of Species*. Darwin didn't have a decent theory of heredity, what we today would call a theory of "genetics."

In fact, even as Darwin was writing, a monk in central Europe, Gregor Mendel, living in what is now the Czech Republic, was doing experiments on pea plants. When his theories were rediscovered at the beginning of the twentieth century, finally biologists had something that could back up natural selection. Although, it wasn't until the 1930s that people realized that Darwinian selection and Mendelian genetics really work together. Before that, a lot of Mendelians thought that change is simply a matter of going straight from one form to another—fox to dog sort of thing, without any need for natural selection. When selection and genetics were brought together, we had what is known as "neo-Darwinism" or the "synthetic theory of evolution"—a new paradigm. After that, of course, along came DNA, molecular biology, and everything really moved forward quickly. But truly, molecular biology filled things out, gave us new techniques and so forth, rather than fundamentally changing things.

Rudge: I wonder if I could just jump in for a second here. I don't want to correct anything Dave has just said, but I think it's important to stress what natural selection is really about. The vital point is that organisms are not just thrown together. They work. They function. They are what the English evolutionist John Maynard Smith called "adaptively complex." The eye and the hand are adaptations that help in the struggle for survival and reproduction. This has an important bearing on our overall theme of science

and religion. The side of theology that is to do with faith and the Bible—
for Catholics especially, tradition, too—is known as "revealed theology."
The side of theology that is to do with reason and physical evidence is
known as "natural theology." The proofs for the existence of God come in
here. Everything has to have a cause. The world is a thing. Therefore, the
world had to have a cause. God!

The most popular proof for the existence of God is known as the "teleo-
logical argument," or "argument from design." The eye is like a telescope.
Telescopes have telescope designers and makers. Therefore, the eye must
have had a designer and maker. The Great Optician in the Sky. God! Darwin bought into this completely. It's true that in the eighteenth century, in
his *Dialogues Concerning Natural Religion*, the Scottish philosopher David
Hume had roughed up the argument, showing that a lot of things are not that
designlike. But nobody took him that seriously. Complex things that work
had to have a designer. It's Murphy's law in reverse. If things can go wrong,
they will. If things are really working well, there must be a reason. Things
like the eye and the hand don't just happen.

Darwin didn't think the designer does things hands-on. He thought the
designer works through unbroken law. By the time he wrote the *Origin* he was
a "deist," thinking that this is the way that God works—he was never an
atheist, although later in life he did become an agnostic—so he was not a
"theist" anymore, that is, someone who thinks that God works through miracles. But, Darwin had read the standard work on the topic. This was *Natural
Theology* by Archdeacon William Paley published in 1802, a year or two before Darwin was born, and he agreed entirely that the eye is like a telescope
and that needs an explanation. This was the point of natural selection.

Davies: I agree entirely. And in fact, I'm glad that you've brought this up
because it does tie into our theme. God's pushed out of science. As Richard
Dawkins has said, only after Darwin was it possible to be an intellectually
fulfilled atheist. That's why I'm firmly in the conflict camp. Science, in!
Religion, out!

Fentiman: Well, hold on a moment. I can see Pastor Hal wanting to get in,
not to mention the Reverend Emily. But, just one quick thing before we
move on to them. You've said that everyone's a supporter of natural selection, a Darwinian. What about people like Stephen Jay Gould? What about
his theory of, what was it called, "punctuated equilibrium"? Didn't he deny
Darwinism?

Wallace: He sure did. It all goes to show that evolution's rotten to the core.

Rudge: Hang on. It doesn't do that at all. Whatever he thought, Gould was always a totally, absolutely committed evolutionist. There was a court case in Arkansas back in 1981, where the Scientific Creationists wanted their stuff taught in schools. Gould was there as a witness for the American Civil Liberties Union speaking up for evolution. Right up to his death in 2002, he was America's biggest evolution booster.

Davies: He was! Gould was a paleontologist and he thought that you should draw a distinction between "microevolution"—short-term moves—and "macroevolution"—evolution over the long scale. He thought that over the long term, you see that evolution goes by jumps or starts. In between there is nothing, or what he called "stasis." That's why he called his theory "punctuated equilibrium." Nothing happens—things are in equilibrium—and then this is broken, punctuated, by quick changes.

As you can imagine, this has been a pretty controversial thesis. Most evolutionists, particularly those who work on short-term changes, for instance, the huge numbers of people who work on *Drosophila*, fruit flies to you and me, think that natural selection does everything. After all, what may seem like a jump in the fossil record could take a huge amount of time even so. In a thousand years, you could breed a fruit fly the size of an elephant, so Gould's jumps could simply be a function of the time scale. Fast in the fossil record is slow in real life. At one point, Gould played with the idea that the changes might in fact be one step—fox into dog—but he jumped back quickly when he was criticized for this. By the way, this kind of thinking is known as "saltationism" from the Latin word, *saltus*, a jump. It was very popular in the years after Darwin.

Rudge: I think it's worth adding just one more point here. This brings us back to adaptation. No modern evolutionist denies the significance of adaptation— the hand and the eyes. But Gould was for many years a strong critic of too much reading in of adaptation. He called this "pan-adaptationism." In a famous article he wrote with his colleague at Harvard, Richard Lewontin, he talked about things called "spandrels." These are the triangular bits at the tops of columns in medieval churches and they are often covered with fancy decorations, like mosaics. But Gould pointed out that really spandrels have no function, they are just by-products of the columns holding up the ceiling. Too often, argued Gould, we think that things are adaptive when they aren't really.

The human chin, for instance, is just the by-product of other parts of the face rather than something in its own right.

Gould called people who think that everything is adaptive "Panglossians," after Dr. Pangloss in Voltaire's *Candide* who thought that everything was happening for the best, no matter how disastrous things may be. He also talked of *Just So* stories, but I guess we'll get to those in our program on humans. The point I want to make here is that Gould's punctuated equilibrium was linked to his attack on too-much adaptationism. He could veer toward saltationism because he was not convinced that all changes had to be adaptive. A strict Darwinian always thinks that change will be gradual because it must stay in adaptive focus.

In fact, it is worth pointing out that Gould's critique of Darwinism was as much metaphysical as something strictly based on empirical evidence. There has always been a divide between biologists—a divide going back to Aristotle—with some thinking that the most important aspect of organisms is their designlike nature, their adaptedness, and some thinking that the most important aspect is their structure, and most particularly the way that structures get repeated between organisms of different kinds. A design person, someone who is known as a "functionalist," is someone who thinks that when for instance you look at the wing of a bird, your first question must be: "What's it for?" The answer of course is: "It's for flying." The structure person, someone who is known as a "formalist," is someone who looks at the wing and asks: "How come the bones of the wing are isomorphic"—what biologists call "homologous"—"to the bones of the limbs of other organisms, like the cat and the porpoise, where the parts are used for very different functions?" The answer today of course is: "They share common ancestry."

The fascinating thing is that these rival visions, as one might call them, transcend the evolutionary divide. Although I don't think the great German philosopher of the late eighteenth century, Immanuel Kant, was an evolutionist, he recognized both aspects of organisms. Then, some of those influenced by him, or thinking in tandem, went the way of functionalism, and some went the way of formalism. For instance, the great French comparative anatomist, Georges Cuvier, at the beginning of the nineteenth century, stressed that organisms are complexly adapted. Conversely, the German poet Johann Wolfgang von Goethe, a man who was intensely interested in science, thought that homology is the key to understanding organisms. Neither of these men were evolutionists—perhaps Goethe became one at the end of his very long life—but they were divided on the function-form issue. Darwin was a functionalist who thought form followed from evolution. His great supporter Thomas Henry Huxley—we've mentioned him before—thought that form was what mattered. He was an evolutionist. Others thought function counted. They, too, were evolutionists. Today, we still have the divide. Dawkins is a functionalist. He says he is somewhere to the right of Archdeacon Paley on adaptation. Gould was a formalist. He thought that homology was the key issue.

Fentiman: Well, I think that's perhaps enough background for the moment. As the audience knows, we have just been listening to Martin Rudge, our professor from Robert Boyle College in Minnesota. We'll get back to you, Martin. For the moment, I want to go over to the Reverend Harold Wallace, our Baptist minister, who I can see is itching to get in.

Wallace: I certainly am! I'm going to disagree with Professor Davies right down the line! But, I hardly feel that I have to make a case. Microevolution, macroevolution, spandrels, *Just So* stories. And that's before we even get to Gould's Marxism! Or the things that haven't really been mentioned, like the origin of life.

Let me make my own position clear first, though. We heard talk, I think in the last program, about Young Earth Creationism, and that's where I stand. Just like 50 percent of Americans. I'm a traditional Christian, a traditional Bible-reading Christian. I think that things happened just like Genesis says. I think that the world and everything in it was created about six thousand years ago. The first person who worked this out from the genealogies in the Bible was an Irish archbishop, Ussher, in the sixteenth century. I think also that the Creation was done in six days and that humans came last. We are made in the image of our creator. I also think that Adam sinned and because of that we were expelled from Eden and everyone fell into sin. I think that sometime later God was really displeased and covered the earth—all of the earth—with water. Only Noah and his family and the animals in the ark survived. As I said last time, I also think that Jesus Christ died on the cross for our sins and made possible our eternal salvation.

But we're only going to get that if we acknowledge Jesus as Lord. It's no good thinking that by being good, we can bribe God. I'm not a strict predestinarian. I think that we ourselves have the free choice to accept God, but we must if we are to go to heaven. "Justified by grace," we call it. I should say, also, that I think the Flood is very important because this was the first of the big disasters to overtake the world. There have been others, like the fall of the Temple. There is going to be another really big one in the future, when the forces of evil start the battle of Armageddon. But some of us will have been raptured up to join Jesus before that.

Matthews: So technically speaking, you're an "Arminian," believing that we humans have a choice to be saved. It's up to us and not rigidly predetermined. Also in the language of theology, you are a "premillennial dispensationalist." You think that there have been periods on earth, dispensations, that are concluded by violent events, and you think that Jesus is going to come to fight in the battle of Armageddon before he sets up his kingdom on earth.

Wallace: Yes, I guess that's right, although the kingdom, in the first place, is only going to last a thousand years, before another battle and the Day of Judgment. I don't know when all of this is going to happen, but it could be in our lifetimes. One big mark, foretold in the book of Revelation, was that the Jews would have to return to Israel before Armageddon. Well, that has happened with the founding of the State of Israel. That's one of the main reasons why we Bible Christians, we evangelicals, are so keen on supporting Israel. It is part of prophecy.

Fentiman: I'm still playing catch-up here a bit, Pastor Hal. Last time you told us that you were not a conflict person. You were more of a dialogue person. You think that science and religion can exist together harmoniously. But it doesn't sound much like harmony to me. You seem to be opposed to just about everything that David Davies accepts.

Wallace: That's true. But, it doesn't mean to say that I'm against dialogue. I think that almost everything that Professor Davies has said is false. But this doesn't mean to say that there isn't a true science that meshes with my religious beliefs. Let's start at the beginning. I don't think you could have a better confirmation of Genesis than the big bang. There was nothing, and then suddenly there was the whole universe exploding into being. Of course, there wasn't nothing. There was Almighty God, but that is part of religion. I have to say that the whole dating issue is about as confused as it could be, except what most scientists say is not right. At the beginning, there were all kinds of perturbations that make it totally unreliable to go back and try to extrapolate as most people do. In any case, most of the dating is inconsistent.

What I really want to talk about, and I'm not surprised that Professor Davies stayed away from it, is the origin of life. It's an absolute mess. There are hundreds of hypotheses and no one has the first clue what they mean or add up to. Notre Dame philosopher Alvin Plantinga—his name has come up before—says (and I quote): "This seems to me for the most part, mere arrogant bluster; given our present state of knowledge, I believe it is vastly less probable, on our present evidence, than is its denial." Even more blunt is Professor Anthony Flew—he is the English philosopher who used to be the archatheist before he found God. He says: "I have been persuaded that it is simply out of the question that the first living matter evolved out of dead matter and then developed into an extraordinarily complicated creature."

The scientists themselves admit this. Take Leslie Orgel, one of the chief origins-of-life investigators. He says it is like a detective story with too many suspects and too many clues and too many solutions. "It would be hard to find two investigators who agree on even the broad outline of the events that occurred so long ago and made possible the subsequent evolution of life in all its variety." He admits that we are probably a long way "from knowing whodunit." I'll say!

Davies: Yes, but he also says: "The only rational certainty is that there will be a solution."

Wallace: That's your old Methodological Naturalism prejudice all over again. There must be a solution. There must be a solution! Of course, there's a solution, but it's not a natural one. God did it directly. Science just can't go that far.

Move on, though. You run into just the same kinds of problems when it comes to all of the major groups. The Cambrian explosion, when all of the really complex organisms suddenly appeared, is an evolutionary scandal. You say it occurred over five hundred million years ago. By your account, life had already been around for three billion years. Where is the evidence for this? There is absolutely nothing! Nothing! You could not have better evidence of God's interventions. Even if you accept a long earth span, which I don't, you have to admit that something happened, something outside the course of nature happened.

However, note something. No one, certainly not us Young Earth Creationists, or Creation-Scientists as we are sometimes called, is denying all change. We are quite happy to accept microevolution. Take something that you Darwinians are always pushing at us. Take those birds that Professor Davies mentioned, the Galapagos finches. There are ten or more different species of finch on the Galapagos Islands in the Pacific. We all accept that they came from a common ancestor. It's just that the change is not from finch to tortoise or something like that. It's from finch to finch. We Creationists insist on some change. No one thinks that Noah's ark carried all of the species we have today. After they left the ark, the "kinds," as we call them, then diversified. It's macroevolution we deny.

Of course, we've seen that you evolutionists are mixed up about this. No one has ever seen a cow turn into a horse. At least Gould was candid about this, although, in fact, he himself admitted that he was a Marxist—he "learnt it at his daddy's knee"—and his belief in rapid change in evolution was no more than a projection of his belief in rapid change in society. And we all know what a vile and broken doctrine Marxism has proven to be.

On top of all of this, everyone knows that natural selection cannot be a very effective cause of change because it is a truism, a tautology. Think of the alternative name—the "survival of the fittest." Who are the fittest? Those that survive! That's true by definition. All of this means that the key mechanism of Darwinism says no more than that those which survive are those which survive! This is hardly the all-powerful basis of a major scientific theory. David Davies has been going on about Karl Popper. I am glad he raised Popper's name, because, in fact, it was Popper more than anyone who argued that Darwinism was not genuine science but a "metaphysical research program." Popper knew what he was talking about. He saw that natural selection is a truism or a tautology. There is no way to falsify it. It cannot be genuine science.

Fortunately, in the last ten years or so a group of scientists have been putting all of our thinking about origins on a new and firmer basis. These are the "Intelligent Design Theorists." They don't go as far as Young Earth Creationists, at least some are young-earthers, but most are not. But, they have been trying to show that major forms of life could not possibly have come through evolution. There must be an Intelligent Designer. The main thinker is a biochemist from Lehigh University in Pennsylvania. His name is Michael Behe and his book is *Darwin's Black Box*. He says that sometimes you get examples of "irreducible complexity," that is, things that could not have come by chance or blind law because they need a number of separate parts and only with all of the parts present and in place will things work. His example is a mousetrap that needs five separate parts—base, spring, and so forth—and will not work without these parts in the right place.

Such an instance of irreducible complexity calls for a designer—a thinking intelligence. If you say otherwise, it's like saying that monkeys striking randomly can type *Hamlet*. It is the same in nature, particularly in the nature revealed by the microscope. Behe talks about the flagellum—a kind of whip-like appendage—that bacteria use to move around. He shows how complex it is and how it would not work without all of the parts. It is irreducibly complex and hence calls for an Intelligent Designer. As it happens, Behe is a Roman Catholic and thinks that the designer is the God of the Gospels, as do almost all other Intelligent Designers. But, the point is that this is not part of his science. He does not want, as a scientist, to say what the Intelligent Designer must be. He just wants to say that as a scientist, he knows that there must be an Intelligent Designer. It is in this sense that we get dialogue. Science points to religion, and religion fills out the gaps left in science.

I should say that I—we—object very strongly to the attempts by evolutionists to keep ID, as we call it, out of the classrooms. We think, like President George W. Bush, that we should teach the issues. We don't want to keep evolution out—anything but. We want it taught. But we think ID should be taught too, so that students can judge for themselves. You shouldn't be criticized on an exam for being pro ID.

Rudge: I just don't know where to begin—

Fentiman: Before you do begin, why don't we let Emily get in a few words. Time to hear from our Episcopalian priest.

Matthews: Thanks, Redvers. As a practicing Christian—and I do want to stress that I am that, for all the jokes that people make about Episcopalians—I simply have to say that I could not be further from Hal. I'm not with David Davies either, but I'm not with Hal at all. For a start, I think it's ridiculous to tie everything into Genesis like he does. It is NOT traditional Christianity to say that we must read the Bible as a literally true work of science. Saint Augustine was absolutely clear about that. Much that we read must be taken allegorically or metaphorically.

Augustine was not an evolutionist, but he thought that things happened developmentally. We heard that it was Augustine who stressed that God lies outside time—he is eternal, he is not just a very old man. Because God is outside time—until he makes himself flesh in the form of Jesus—for him the thought of creation, the act of creation, and the product of creation are as one. Augustine believed God created seeds of life with potentialities that then sprang into life and full being. Not evolutionary as such, but certainly something that could be evolutionary.

You see, for me, this whole business of reading the Bible in the way that you do is to miss entirely the proper messages. Take Noah and the Flood. I think this has nothing at all to do with aquatics or whatever. Nor is it a lesson in zookeeping. How many sons did Noah need to muck out the elephants and the other large mammals each day? I think as much as anything else it is a story about the futility of simplistic solutions. God sees the world is in a mess, so he cleans it right out, except for Noah and family. But then what happens? Noah gets drunk and one of his kids makes fun of the old man while he's lying there naked and out of it. It seems like sin has not gone after all. That for me is the real message—at least, one of the real messages—of Noah. Going at a serious problem and thinking that all you need to do is wipe everything out and start again is basically not going to solve your problem. This for me is why the Bible is such a living document. It has nothing to do with geology and certainly has nothing to do with all of that stuff about Armageddon and Jesus coming again to do all of that fighting.

Rudge: Perhaps I can just put in a footnote on all of this. Emily is absolutely right that Young Earth Creationism has little to do with traditional Christianity. The top scholar on all of this, Ron Numbers at the University of Wisconsin, has shown that it goes back to the beliefs of the Seventh-day Adventists, the group started in the middle of the nineteenth century by Ellen G. White. Because she wanted to make so much of the Sabbath, it was she who stressed the literalness of the seven days of creation and so forth. No one else bought into that sort of thing.

We've all heard of the Scopes Monkey Trial, when a young schoolteacher was prosecuted in Tennessee in 1925 for teaching evolution. He was defended by the best-known attorney of the day, a real agnostic, Clarence Darrow, who had just saved Leopold and Loeb, the child killers, from the electric chair. He was prosecuted by the politician William Jennings Bryan. I guess most of us have seen the fictionalized version, the play or film *Inherit the Wind*. The Bryan figure is presented as a buffoon. Matthew Harrison Brady, who takes the stand about the Bible, is shown to be completely confused. In fact, the real Bryan was quite clear that he did not think the days of creation were necessarily six literal days and he did not subscribe to such a short earth span.

It wasn't really until the 1960s that Young Earth Creationism became so fashionable as to become the norm. This was when a Bible scholar named John Whitcomb teamed up with Henry Morris, an engineer—a guy who had deliberately trained as a hydraulic engineer, so he knew about these sorts of things—to write a book that promoted the short earth span. *Genesis Flood*, as they called their book, was a runaway success, and the rest—as they say—is history. A very short history!

Fentiman: Well, after that VERY lengthy footnote, perhaps we can get back to Emily.

Matthews: Thank you. As I have said, I think that Hal is reading the Bible quite incorrectly and nontraditionally. I should say, also, that I don't buy into all of his supposed tolerance about a dialogue between science and religion. I think that Scientific Creationism and Intelligent Design Theory—what I call Creationism lite—are just crude, biblical literalisms—what we used to call Fundamentalism, and I'm sure that Martin can and will give us a footnote on this, too—and Hal and everyone else really knows that these are not science at all. Evangelicals like Hal pretend that they are science, so that they can get around the U.S. Constitution. This separates science and religion and tells us that you cannot teach religion in state schools. So the Fundies pretend that their stuff is really science, rather than Genesis, and claim it can be taught in schools. Thank God—and I mean thank GOD—the courts have so far seen around this trick.

Davies: Yes, let me push in here. The monkeys and typing-Shakespeare story is just silly. Richard Dawkins showed that you can program a computer to produce random letters and so long as the earlier successes are kept and not thrown away, sense—Shakespeare's lines—can occur in fewer than thirty moves. Dawkins showed how you can get "METHINKS IT IS LIKE A WEASEL" from a nonsense-line like "KDS EODK DIAPOEOAPAP LCKJAJE."

Wallace: It's not a fair analogy. You know that "METHINKS IT IS LIKE A WEASEL" is part of Shakespeare before you start. Nobody knows—at least nobody knows according to you evolutionists—that something like Tyrannosaurus rex was going to appear before you started.

Rudge: Point well taken. What Dawkins does show, however, is that random processes can generate meaning, so long as they are cumulative. The trouble with a monkey striking out Shakespeare is that it is going to be random all of the way. What Dawkins shows is that if have a system where if you get something right, it is preserved, then randomness can give you what you want. Suppose you have a system where it is all or nothing. Then "METHINKS" would take ages. But if you have a system where if you get one letter right, then it sticks and then you go on until you get another letter right, "ME-THINKS" can come out pretty quickly. Hal is right that this on its own doesn't give you natural selection, but it isn't really intended to.

Davies: You're right, it isn't. My main point, however, is that all of this stuff about "balanced treatment" or "teach the issues" is nonsense. You just don't teach balanced treatment in schools or universities. You certainly don't do it because some part of society thinks some silly idea is the truth. The Christian

Scientists have all sorts of crazy ideas about medicine. I don't want this taught at the medical school on my campus. I want the doctors to know about these things, because they will certainly encounter them in practice. But I don't want them to believe them.

Wallace: Yes, but we're not just a minority. We are the majority in America.

Davies: That cuts no ice with me. The majority of Americans in the South in the fifties thought that African Americans should not use their washrooms. They were wrong.

Wallace: There you go again being offensive rather than arguing.

Matthews: Fellas, back to the topic! Dave does hit on a good point. Most whites in the South at the time of the Civil War used the Bible to justify slavery. Saint Paul told the escaped slave to return to his master. There was no emancipation declaration there! Of course, I think a biblical case can be made for the abolition of slavery. You are hardly following the Sermon on the Mount if you have slaves. The main thing, however, is that simply quoting the Bible without understanding or interpretation is bad for science and bad for morals. I no more want to follow Saint Paul on slaves than Genesis on origins.

Fentiman: Yes, but can we swing back to the science and religion issue? Emily, you've said that you're an integrationist, but how is that in the evolution case? We know why you don't go along with Pastor Hal, but why integration rather than conflict like David Davies?

Matthews: Good question. I accept natural selection. Of course I do. I am not sure, however, that I am as enthusiastic about it as Professor Davies is. Like Stephen Jay Gould, I am inclined to think that the tight designlike nature of the world is a bit of a myth. Note that far from being a refutation of the kind of God I believe in, it is just what I would expect. My God is not the god of the philosophers. My God is working with us, helping to shape existence as a cocreator. I expect things to go wrong or to be rough at the edges. That is what a state of becoming rather than a state of being requires. So, in that sense, I am very much an integrationist. My science and my theology are as one. But, note that I am not doing traditional natural theology. I am not proving God from nature. I am saying that my view of nature and my view of God are one and the same. On this whole matter,

I am very much in agreement with the greatest British church figure since the Reformation, from the nineteenth century, John Henry Newman, who started life as an Anglican and then converted to Roman Catholicism and ended as a cardinal.

Wallace: I think you might want to make a case for the eighteenth-century founder of Methodism, the great evangelical John Wesley. It is true that his supporters broke away from the Anglican Church, but still he was a towering figure.

Matthews: Agreed! I would not for the world downplay Wesley or his brother Charles, who wrote all of those wonderful hymns. But Newman is my hero, and nowhere more than when he wrote to a correspondent about his seminal philosophical work, *A Grammar of Assent*: "I have not insisted on the argument from *design*, because I am writing for the nineteenth century, by which, as represented by its philosophers, design is not admitted as proved. And to tell the truth, though I should not wish to preach on the subject, for 40 years, I have been unable to see the logical force of the argument myself. I believe in design because I believe in God; not in a God because I see design." Continuing: "Design teaches me power, skill, and goodness—not sanctity, not mercy, not a future judgment, which three are of the essence of religion."

Fentiman: Can we move along, please? I'm keen to get Martin Rudge into the conversation before we finish this program. I can see that you're itching to go, so over to you!

Rudge: Thanks. For a start, let me say that I really am quite surprised that Emily would be quoting John Henry Newman in support of her integrationist position. It seems to me that he is much more inclined to my separatist position. Newman says don't look to science for design, and, yet, this is just what Emily seems to be doing. Hal is doing this, too, but he is up front about it. I agree that it is open for a religious person to say: "I don't think that you can go from the design of the world to God, but as a Christian or whatever, I see the glory of God in the creation." This is what Newman is saying. Emily is saying something much stronger, along the lines of: "The world is so well designed that we have to accept God." I don't think this is true, but true or false, it is not Newman's position.

I might add I am also a bit surprised that Emily brings up Augustine with quite so much gusto. It is true that he thinks that God is outside time, but it seems to me that the whole point of Process Philosophy—the viewpoint based on the thinking of Alfred North Whitehead—is that God is brought right into time. He—or she or it—is now working alongside humans in the act of creation. This is a major reason why many mainstream theologians want nothing to do with Whitehead. I agree with them. I just don't think it has much to do with Christianity at all. Emily is right that ideas change and that applies to religion and theology as much as to anything else. But there comes a point when the change is so much that the original is gone. I think humans evolved from single-celled organisms like bacteria, but that doesn't mean that we are really bacteria.

Of course, for all of our disagreements, I am still closer to Emily than to Hal. Turning to him, let me start with the charge that natural selection is a tautology and that consequently Darwinism cannot be a genuine theory. There just has to be something wrong with this objection, if only because it is a major part of modern evolutionary biology that the fittest do not always survive and reproduce! An important theoretical biologist of the twentieth century, Sewall Wright at the University of Chicago, showed that sometimes in small populations the vagaries of breeding mean that the less well-adapted, what are known as the less fit, can predominate over the more well-adapted, what are known as the more fit. He called this "genetic drift." Its significance has been much debated, but no one thinks that it is a contradiction—which it would have to be, if the opposite were a tautology.

It is right that fitness has to be defined in terms of success in surviving and breeding. But there are some empirical claims here—claims that could be false. No one is simply saying that fitness is reproductive success and that reproductive success is fitness, and just leaving it at that. First, natural selection is drawing attention to the fact that there is a differential survival and reproduction going on. It could be that every organism just budded off an identical twin and then died. There would be no struggle or selection at all. The claim is that organisms are different—could be false, although it isn't—that some get through and some don't—could be false, but it isn't—and finally that the reason for success or failure is, on average, linked with the different features possessed by different organisms. Having eyes and ears, rather than being blind and deaf, really does count. Second, linked with this, there is the assumption that what works in one case works in another. This is a kind of inductive inference, and perhaps because Popper disliked induction so much, he tarred selection with the same brush. But whatever Popper thought, the fact is that Darwinians think that if, let us say, black skin is of adaptive value in one case, let us say as camouflage against a black background, then it will be of adaptive value in other cases too. This could be false, in which case selection would fail. Indeed, the whole point of genetic drift, because the claim is that it doesn't always hold!

So the simple fact is that selection is far from being a tautology. Although, in a way, the problem is a bit more complex than I have suggested so far. It is true that scientists work within the context of major theories—the philosopher-historian Thomas Kuhn called them "paradigms." So for instance, a physicist might be working against the context of Einstein's relativity theory. In biology, we work against Darwinian evolutionary theory. But the day-to-day work of the scientist is a bit more prosaic or mundane. One doesn't spend all of the time quoting the major theory. One deals with little problems of theory or experience. One tries to explain something puzzling, like why a particular animal has a higher blood pressure than most or some such thing. To do this, one builds limited little theoretical models—pictures that try to explain certain facets of experience. Earlier in the program, David Davies gave us an example of this in practice when he made up his little example about the various offspring, their colors, and their reproductive successes.

Often these models are entirely done on spec, as it were. You pretend you have a situation, and then see what happens. What occurs when two predators put different selection pressures on prey, for instance? At this level, everything is theoretical—nothing is empirical—and so, in a way, you assign selection values by fiat, as it were. By definition in the model, organisms of type A lose 20 percent of their offspring in a generation. This is if you like tautological. But, then you have to see if your model applies to the real world, and here the empirical factor—the falsifiable factor, in Popper's terminology—comes into play. So ultimately, nothing is being done just by definition.

Fentiman: So much for natural selection. What about Intelligent Design? Professor Davies, you are the scientist, and so I guess that, in a way, it is your thinking most under threat. What do you have to say on the topic?

Davies: A lot, and it's all negative! The Intelligent Design claim is that the organic world shows irreducible complexity. Behe must be a brilliant teacher. I certainly give him that much. He is so good at picking up analogies to explain complex points. To illustrate irreducible complexity, he focuses on a mousetrap. He points out that it has five parts: the spring, the snap, the base, and so forth. Each of those parts is necessary and if they are not all perfectly in place, nothing works. His conclusion is that something like this could not have been produced by blind law, especially not by natural selection. Natural selection is gradual and mousetrap functioning precludes this.

Nice example, but it doesn't really work! Start with the fact that irreducible complexity is a bit of a mushy notion and that, in any case, a mousetrap does not seem to be irreducibly complex. You can make a trap with fewer than five components, starting with eliminating the base and bolting the trap to the floor. You can cut down on other parts also, for instance by combing spring and snapper. On top of this, it is very improbable that Behe's examples are as irreducibly complex as he claims. One of his favorite examples is of the so-called blood-clotting cascade. When you cut yourself, the blood soon stops flowing and clots—unless you are unfortunate enough to be a hemophiliac, in which case, you stand in danger of bleeding to death. It turns out that blood clotting is a rather complicated phenomenon involving, I think, about twenty-nine different steps in order—hence, the term "cascade." Behe argues that if you take one of these steps out of the line, then the whole process collapses. According to him, it is twenty-nine or nothing, and twenty-nine parts all put in place at once is impossible through natural selection.

The experts on blood clotting, particularly a scientist in California called Russell Doolittle, say that that is nonsense. You can get blood clotting with fewer steps, and you see this in animals lower than humans, which is what you would expect if evolution is true. In any case, the parts are not all separate, but often modifications of an earlier step. So, it is quite plausible to think that blood clotting comes about through the multiplication of a process that starts off doing the job less well and then improves over time with the multiplication.

More generally, we find that complex organic systems do not necessarily have to come into being in one fell swoop or not at all. Usually, such systems are cobbled together from parts that are already functioning in the system. Then, they are recycled for other uses. A good example is the Krebs cycle, a highly complex process with many steps, used by the cell to provide energy. It did not just spring into being. It was a "bricolage," built bit by bit from other pieces. I have a quote here: "The Krebs cycle was built through the process that [the Nobel Prize winner François Jacob] called 'evolution by molecular tinkering,' stating that evolution does not produce novelties from scratch: It works on what already exists. The most novel result of our analysis is seeing how, with minimal new material, evolution created the most important pathway of metabolism, achieving the best chemically possible design. In this case, a chemical engineer who was looking for the best design of the process could not have found a better design than the cycle which works in living cells."

You see, in a way—and I guess this is my really basic point—people like Behe have simply got the wrong end of the stick. They think that if you have something functioning well, like the bacterial flagellum or the blood-clotting cascade, then, this is how it always must have been, functioning on its own from the beginning. But, that's just not how evolution works. It is a fumbling process. And often, you cannot tell the process just by looking at the product. Take an example—if Behe can do this, so can I—of a stone bridge built without mortar and held up simply by the stones pushing against each other. If you tried just to build the bridge, starting on both sides and then moving inward, before long the sides would collapse and you would never meet in the middle. What you must do first is build an embankment or a trestle, and then lay the stones on it. When they are in place, with a keystone in the middle, you can remove the supports and the bridge stays in place—similarly, in evolution. You might have a process that works nicely. Then other parts piggyback on the process until they join up and start functioning. Then the original process drops out. You could not get the new process on its own, but you can in just the same manner as building a bridge.

Counterpoints aside, what is really offensive about Intelligent Design is that it is a science stopper. It says: "Give up, there is no solution." No scientist would say that unless there were a hidden religious motivation. Blood clotting is a challenge—at least, it was a challenge before it was cracked—not a reason to turn to God. Again and again in science, something that seems impossible to one person or even a generation proves soluble to another person or generation. Sometimes you need new theories, like the double helix; sometimes you need new tools, like computers, but that is the game of science. Good science has more problems at the end of the day than at the beginning. That is what makes it so exciting.

Wallace: I suspect that we shall just have to agree to disagree. But, there is something I want to pick up on before we finish tonight. Everyone is having a crack at me, so I want in return to have a crack at you—at Professor Rudge, in particular. I am a bit worried. Go back to what you were saying about form and function. I don't want to challenge your history. I am sure that you know much more about it than I do. I don't, at this point, want to challenge what you said about evolution. Even though I know that evolution is not true, I am quite happy to accept that Goethe became an evolutionist and that Darwin and Huxley were evolutionists, and the same is true for those who came later. My worry is about what you seem to be saying, implying at least, about science. You seem to be saying that we have had these two traditions, formalists and functionalists, and the facts really don't matter. You fall on one side or the other of these traditions with about as much justification for preferring maple walnut ice cream over chocolate ice cream. Nothing is going to change your mind about the ice cream flavor and nothing is going to change your mind about form or function.

This, it seems to me, makes science absolutely subjective—really, just a matter of background. If you grow up in Canada, you will like maple walnut ice cream. If you grow up in Belgium or Switzerland, you will like chocolate ice cream. It has nothing to do with rationality or objectivity—the same for form and function, rival perspectives and no independent way of choosing. As a Christian, that's certainly not acceptable to me. I believe that we have God-given powers to discover his creation and that he is not going to let us be deceived. There is a real world and one unique description of it. Evolution is false. That is not a matter of choice or opinion.

Davies: Funnily enough, I agree with Hal on this one. As a Popperian, I don't think we can ever get to the absolute truth, but I do believe that there is a real world, independent of human perception or thought, and there is one proper description and explanation. Science is a way of approaching it, asymptotically, as it were, getting ever closer, even if we never get right there. Evolution is true. And that is not a matter of choice or opinion!

Matthews: I, on the other hand, am against both of you and happily and readily endorse what Martin has been saying! I just don't like these sharp divisions between me and you, us and them, objective and subjective, even God and his creation. I recognize that religion is cultural. What was acceptable at the time of Paul, for instance, with respect to women and slaves and gays, is no longer acceptable today. We have to put things in a historical context, in a human context. Why should science be any different? In respects, it is what the philosophers call a "social construction," something that sits on or in the society in which it is formed and used and cherished. As an integrationist, I expect science and religion to be of the same logical type, and so I have little time for the objectivity supposed by either Dave or Hal.

Rudge: Since I seem to have got us into this mess, perhaps I can try to dig us out. We have talked about metaphor already—the organic metaphor and the clock metaphor—and I would want to push this insight. Think of the metaphors of Darwinism: struggle for existence, natural selection, tree of life. Since then, we have had the American population geneticist Sewall Wright claiming that organisms seem located on what he called an "adaptive landscape," with the most adapted on the peaks and the others in the valleys; Richard Dawkins is famous for talking about "selfish genes"; and often today, evolutionists say that competing lines of organisms, predators getting faster and prey getting faster in response, are being caught up in "arms races." Some philosophers have argued that these metaphors are only useful for discovery of new facts; they have a heuristic value, and in the mature science they drop away. I can't say I see much evidence of that. Natural selection is as alive and well today as it was when Darwin first proposed it publicly in 1859, in his *Origin of Species*.

It seems to me that these metaphors all suggest that evolutionary thinking is a child of its culture. Both Darwin and today's evolutionists talk about organisms displaying a "division of labor." The famous Harvard entomologist Edward O. Wilson, the world's leading expert on ants, argues that the different castes—foragers and nursery workers and soldiers—display just such a division of labor. A soldier is huge and fierce and could not look after the young very efficiently, and a nursery worker is small and no good at fighting. This metaphor is right out of the eighteenth century when the Scottish economist Adam Smith argued that to make pins, it is far better that one man do one task, head making for instance, and another man do another task, point making for instance.

So, I would argue that you just can't take science out of culture. But, I don't think it means that empirical testing is any less important. If natural selection didn't lead to any good predictions in the real world, we would soon drop it. The same is true of the division of labor. If Wilson found that the ratios of his foragers to soldiers did not turn out as predicted, he would abandon the metaphor of division immediately. I think Popper is right. We edge closer and closer to pictures of reality. The only thing is that there are different pictures. Some are better than others. Those that are not very good at all just get dropped. Sometimes, perhaps form and function is a case in point, you are always going to get some people who will support one or the other.

Fentiman: I don't quite see where the metaphors are here.

Rudge: Function is easy. You are looking at the eye as if it were a telescope. The heart as if it were a pump. The organism as a product of design. Form? Well, perhaps Kant gives a clue. He talks about snowflakes, with their perfect hexagonal shapes—six equal, identical parts, radiating out from the center—as the perfect example of form. Perhaps we are looking at things, living things, as if they were crystals or some such thing. Always repeating. Perhaps like snowflakes, no two exactly the same. Formalists often point to the repetitions within organisms—the repeating backbone, the repeating parts in phyllotaxis, as well as repetitions across organisms. Snowflakes again.

Davies: This is all very well, but how do you tie in what you are saying to science and religion? For me, science is something testable. Religion is not.

Rudge: I'm not saying that science is not testable. Anything but! What I am saying is that there is culture in science. In fact, in the case of evolution I would go further. I could imagine a race of intelligent beings on another planet, a race of intelligent science users, who did not have any idea of evolution. I do not suggest that they would be Creationists, because that is our thinking but wrong. Rather, I suggest that they might simply not cut up the empirical pie as we do. Evolution is a child of Judeo-Christianity. It is an answer to the question: "Where did we come from?" This is a question posed by the Old Testament. The ancient Greeks did not have this sense of history. For them—for Plato and Aristotle—the universe was eternal, without beginning or end. It was not so much a question of being against evolution as that evolution did not answer any questions they were asking. So my extraterrestrials might know a lot about organisms, but simply not put things together as we do. Why would you think in terms of a tree of life, without Genesis? Perhaps they would think more of a directionless web or mesh or something.

Davies: Actually, we now know that genes can be transferred via viruses across big gaps, from and to organisms that are very different. Nature was into genetically modified organisms long before we were. A lot of today's evolutionists do think in terms of nets.

Fentiman: But what's the relevance to science and religion?

Rudge: I think simply that even those of us who are nonbelievers should be a little more modest in what we say. It simply isn't the case that science is something that is pure and objective and culture free. And that religion is impure and subjective and culture laden. You cannot just praise the one and condemn the other on those grounds. They are both human constructions in some sense. As you know, I think that science and religion address different questions, but that is another matter. I also think that the metaphors of science do provide certain constraints, and perhaps this does leave open a place for religion for those who want it. Let me explain.

Fentiman: Not right now, I am afraid. We are right out of time. This is the host of *Eternal Questions*, Redvers Fentiman, wishing you a good night. Join us next week!

~

CHAPTER THREE

Program Three: Problems

Fentiman: Good evening, again. Welcome to the third of our programs on science and religion in the *Eternal Questions* series. Our usual panel is back with us. On my extreme left is David Davies of the Massachusetts Institute of Science. He is taking the position that science and religion are at war. On my near left is Martin Rudge of Robert Boyle College in Minnesota. He thinks that science and religion speak to different kinds of issues. On my near right is Emily Matthews, an Episcopal priest here in town. She thinks that science and religion can be integrated. And on my far right, last, but not least, is the Reverend Harold Wallace of Atlanta. Although he rejects much of modern science, he thinks that science and religion can speak to each other in dialogue.

Wallace: I don't reject much of modern science. I reject those parts of modern science that are wrong.

Fentiman: Fair enough. Now, we have a lot of issues left over from last time. A real gallimaufry, as one might say.

Davies: A real what?!

Fentiman: A real gallimaufry. It means a hotchpotch.

Davies: Then why didn't you say so?

Fentiman: I did! It is just that I used a word that you don't know. But from now on, if you insist, I'll keep it down to four-letter Anglo-Saxon words. More seriously, last week's discussion about evolution really only started into the issues. This week we should continue with them. The only real condition that I want to impose is that we put humans on one side. We will be talking about them next time, and in our final program. I'd like to start us off on the topic of the origin of life. Given what the Reverend Harold Wallace said last time, this is clearly a topic that we should be tackling. Then, I really think we should pick up on the question of alternative theories to Darwinism. We mentioned Gould and punctuated equilibrium in the last program, but I wonder if we have fully exhausted that issue. I keep coming across terms like "self-organization" and I would like to know a bit more about these and whether they tell us something about evolution. In particular, whether they tell us something about evolution that is pertinent to the question of science and religion.

Emily, I can see you nodding your head right now, and David and Martin, I can see both of you shaking your heads. So, I guess I'm on to something! Finally, assuming we have time, I'd like to branch out a bit. We have been focusing on biology, but I do want to ask if there are issues we are ignoring in the other sciences. We mentioned last time that the focus of the science-religion relationship used to be on physics and now it seems to have shifted to the biological sciences. I wonder if this is entirely true.

Let us turn to the origin-of-life question. What about the challenge flung down last time? What was it that the former atheist Tony Flew said? "I have been persuaded that it is simply out of the question that the first living matter evolved out of dead matter and then developed into an extraordinarily complicated creature." Hal Wallace, I will let you have the first word, if you want to repeat or elaborate on your challenge.

Wallace: Thank you. As I said last time, I regard this as a real stumbling block to the naturalism program. You can't go from apples to oranges. Even more, you can't go from the nonliving, from inert chemicals, to the living, not even the simplest life forms. Francis Crick, the codiscoverer of the double helix, admitted this. He proposed the hypothesis of panspermia, the idea that life here on earth was seeded by forms from outer space. But, this only puts the problem off for the naturalist. If not here on earth, then where? The Bible says that God created life miraculously, and truly that is the only sensible conclusion. Spontaneous generation—instant life from mud or ponds or whatever—is just plain false. The French scientist Louis Pasteur showed that in the middle of the nineteenth century. You just have to go with miracles, and for me the early chapters of Genesis say all that is needed to be said.

Matthews: Obviously, I don't agree with Hal that a simple miracle will do the trick. I don't buy into his simplistic reading of Genesis. However, I must say that I do have some sympathy for his thinking. As you know, I want to stay on board with science, but I, too, am not entirely convinced that science as we know it today will do the trick. Life is so incredibly intricate, so well put together or organized. Paul Davies, the astronomer who wrote *God and the New Physics*, says that we need new laws to explain the origin of life—new laws, the sense of which we can still only grasp inadequately. Like me, he denies that life's beginning was supernatural, but—again like me—he does want to leave open the possibility that there was something more. "I do believe that we live in a bio-friendly universe of a stunningly ingenious character." He says that he is looking for "new philosophical principles" which will have "immense philosophical ramifications," and seems to think that information will be important. I couldn't agree more. When we get to talk about things like self-organization, perhaps I can sketch out where I think we should be going.

Fentiman: Now, who on the other side is going to pick up this challenge?

Rudge: Let me start, because I really think that this is as much a philosophical question as one of science. Then I'll hand over to David Davies.

The philosophical side to the question is about the very nature of life itself. What is it that we are talking about? What is life, that distinguishes it from nonlife and whose origin is such a puzzle? The traditional answer is that it is some *thing*. It is an entity, if not quite like a chair or a table, then perhaps like an electric current or a beam of light. This sort of thinking goes back to Aristotle, some three or four centuries before Jesus. It's amazing how much does go back to Aristotle; although, when you are talking about organisms, perhaps that's not so very surprising, given that he was a first-class biologist, as well as a philosopher.

Aristotle's thought that life is a thing is not stupid. What's the difference between a dead cow and a live cow? The live cow has a life force that the dead cow doesn't have. Remember those old *Topper* movies about the couple killed in a car crash, who then came back as spirits. When they died, their life forces seemed to rise up above the wreckage. That is the sort of way that tradition characterizes life. The trouble with this kind of thinking is not that, in real life outside the movie theater, we don't ever see the life force—no one has ever directly seen the DNA molecule—but that it doesn't seem very helpful. Once you talk about the DNA molecule and how it's a double helix and so forth, you are off and running. There are masses of bits of science that you can do. Life forces, to the contrary, don't do much at all. What help is it to be told that a live cow has a life force? You seem to be able to do all the studies that you want on cows, feeding behavior for instance, without once mentioning the force.

It's worth mentioning that around 1900, there was a group of people called "vitalists"—a prominent member was the French philosopher Henri Bergson—who thought that what made for life was indeed a force. Bergson called it the *élan vital*, but soon people rejected it, pretty much once and for all. One major reason for the rejection is that not long after the vitalists starting pushing the life force, thus bringing the whole issue to the front of people's thinking, several folks, including the British biochemist J. B. S. Haldane and the Russian scientist Alexander Oparin, responded by arguing that the Aristotelian tradition had got the problem mixed up. The very quest for a thing that characterizes the living is a mistake. The distinctive point about life is that it is organized—molecules are put together in such a way that they sustain an organism's eating and drinking and copulating and so forth. There is no mystery. A cow in the slaughterhouse with a bullet through its brain has its parts mixed up and cannot function anymore.

Sorry, I am starting to slip into lecture mode, but the information is relevant for our discussion. All living things, including humans, are living because they are organized in a certain way. There is not some special life force that makes them this way.

Wallace: If I can interject, it seems to me that you are jumping from the frying pan of vitalism and into the fire of Marxism. Both Haldane and Oparin were notorious Marxists, and their philosophy was just that of Marxism. More accurately, it was the philosophy of Marx's sidekick Friedrich Engels who argued that organization is the key to understanding the world.

Rudge: That is absolutely true about Engels, although in this case, it is irrelevant. Both Haldane and Oparin came up with their thinking about life before they became Marxists. In any case, as General William Booth, the founder of the Salvation Army, said when critics complained that his hymn tunes were pinched from secular sources, why should the devil have all of the good tunes? Why should the Marxists be able to pinch all of the good ideas for themselves? I'm all in favor of trains running on time, even though this seems to have been the one real achievement of the Fascist dictator of Italy, Benito Mussolini. The point of relevance here is that the origin-of-life problem is not one of looking for a new substance. It is a question of organization. What makes something function, and how does it reproduce itself?

In the old days, people used to speak of the "spontaneous generation" of life from inorganic matter—worms appearing in ponds when lightning struck and that sort of thing. Today, we think that these old questions of spontaneous generation are not so much unanswered but unanswerable. People who cast the origin-of-life question in these terms were looking in the wrong direction. Now, obviously, thanks to molecular biology, we do know an awful lot more about the organization of life. We know that the old-fashioned Mendelian genes are truly long molecules. They are deoxyribonucleic acid, or DNA, twisted around another in a double helix—this was the great discovery of James Watson and Francis Crick in 1953—and that the order of their parts is how they carry the information for building life forms. We know also that this information is taken by another molecule, ribonucleic acid, RNA, and used to pick up building blocks, the so-called amino acids. How then these amino acids are linked up to make proteins, the things that build the cells and that act as facilitators—the technical term is "catalysts"—to make the processes in the cell run smoothly.

Now also, thanks to the bringing together of embryology and molecular biology, so-called evolutionary development, or evo-devo for short, we are learning how the organism develops. For instance, we have learned that life does not start anew with each different kind of organism, but is built much more on a kind of Lego pattern. There are chunks of working parts that are used again and again in different ways—different organized patterns—to make different organisms. Humans and fruit flies share similar strands of DNA and the consequent proteins! With the same bits of Lego, you can build a spaceship or the White House. With the same bits of organic molecule structures, you can build a classy primate or an insect!

Of course, this in itself doesn't answer the question of how things get going. But, now we have a much better idea of what we want to get going. And, while this doesn't speak to all of Hal's worries, I think it goes some way to answering Emily's worries. Life is mysterious, in the sense that a great Bach cantata is mysterious. You want to stand back and wonder at the glory of it all. How can this be? But, it is not mysterious in the sense of being magical. It all works according to the usual laws, just like Bach works because of the usual laws or rules of music. In fact, Bach works precisely because of the laws and so does life. You don't need some extra law to explain how RNA copies information from DNA. For both Bach and life, the wonder is in the organization, not the laws.

Matthews: But, isn't that precisely my point! How do you get the organization in the first place? Bach was so exceptional. So, also, is life!

Fentiman: It is time for Professor Davies to get into the act.

Davies: It's best, I think, to divide the topic up into a number of stages. The first is getting the very building blocks of life, particularly the amino acids, those smallish molecules that Martin correctly just told us are linked up to make proteins, the constituent parts of cells, as well as the catalysts that drive the processes. Also, one needs the components of the nucleic acids that program everything. As it happens, this seems to have been the easy part. In 1953—the same year as the double helix—the grad student Stanley Miller, working in Chicago in the lab of Harold Urey, simulated the conditions they thought would be on earth just before life—compounds in warm ponds, sort of thing—and then shot electricity through the mix, as though it was lightning. He didn't get worms, but he did end up with amino acids, naturally formed as it were. Others later did the same for the bits of the nucleic acids.

So, there was a fast start to the origin-of-life question. But then the troubles started to multiply. How do you get the bits linked up so that they function, especially after you have made such a fuss and to-do about organization? You have the classic chicken-and-egg situation. No chicken, no egg. No egg, no chicken. No DNA, no proteins. No proteins, no DNA. Recently, however, things have started to move forward again. The key seems to be RNA. It has long been known that in some organisms it carries the genetic information—there is no DNA—and, thanks to great discoveries, it is now known that it can serve as its own catalyst and get things going. Of course, you still have the problem of getting RNA in the first place, but apparently sometimes the molecule can link up on inorganic clays, and so can be formed naturally. There are lots of experiments showing that, once you get RNA, it can mutate and a very strong selective process can take over, picking out the best reproducers and so forth.

Things are not that blind.

Wallace: Yes, but what about the fact that now many people don't think that the world was like Miller and Urey supposed. They don't think the amino acids could have been formed, or if formed that they would have been stable?

Davies: I am glad you asked that question. You are absolutely right that to-day a lot of people think that the origin of life did not occur in some little warm pond—an idea, incidentally, that Charles Darwin once hypothesized in a letter to a friend. But there is another candidate! It is more than likely that life originated from the materials forced out of deep sea vents. I'm talking about the things produced by continental drift as the new plates are pushed up from below. There are lots of minerals down there, and the heat provides an energy source. So your objection is nothing like as frightening as it would have been once. Not that I want to minimize the problems that still bedevil the origin-of-life question. We have still got to put everything in the shell of the cell, for instance. This is a problem that Oparin worked on for much of his life. And then you have got to keep moving in, providing the various parts of the sophisticated cell. We've mentioned Lynn Margulis be-fore, and I'm sure that her major contribution will come up later, when we turn to the history of early life here on Earth.

Wallace: Yes, but it's all guesswork and surmise. Hypothesis on hypothesis. You claim to be a follower of Sir Karl Popper. He said it all: "An impenetra-ble barrier to science and a residue to all attempts to reduce biology to chem-istry and physics."

Davies: I am a Popperian about methodology. I often disagree with him about science itself. We've heard that he said that Darwinism was not gen-uine science, but a metaphysical research program. I think that is just stupid. J. B. S. Haldane, as always, had a pithy response to that kind of argument. He said something like: "Show me a fossil cow in the pre-Cambrian, and Dar-winism is false." He meant show me a mammal long before mammals were supposed to have evolved and Darwinian evolution through natural selection cannot explain it—more than that, positively denies it. There are all sorts of other things that could falsify Darwinism. Finding an island off the west coast of Africa, say, and then discovering that all of the plants and animals looked Australasian, not African, would be another refutation. The fact that you don't find these sorts of things doesn't mean that Darwinism is unfalsifiable. It means that it is unfalsified. It could be true!

My point is the following. It has only been fifty years since we have known about the double helix. In that time, we have gained a huge amount of in-formation about the workings of life. At the same time, admittedly, we have learned that the origin-of-life questions are hugely complex. What then is the right position to take? Is it to put it all down to miracles, like Pastor Hal? Or is it to call for special laws, like the Reverend Emily? Or is it to say that it is silly to give up the hunt when, only now, finally, do we have the tools to do the job? Is it to say that only now when—as Martin Rudge was pointing out—we know that we are not hunting for a substance, but that it is all in the organization that we are truly on the right track? For myself, I see Nobel Prizes in the future. I think the origin-of-life question will soon be one more nail in the coffin of religion.

Fentiman: Let's keep on our time schedule. Next, I want to turn to challenges to Darwinism and how they might affect the science-religion relationship. Emily, I can see that you are eager to jump in, so why don't we let you have first crack at this.

Matthews: What I want to say really follows on what I was saying last time. Like Stephen Jay Gould, I am not so sold on the Darwinian picture. In the world of organisms, a lot of people now are starting to suggest that nature itself—the laws of physics and chemistry—can truly put in order, as well as take it away. People like the American theoretical biologist Stuart Kauffman talk about "self-organization," or rather cleverly "order for free." Kauffman follows in the tradition of the early twentieth-century Scottish morphologist D'Arcy Wentworth Thompson, who thought that natural selection was overrated and that much of the order in the living world is due to the way that the basic laws work. Thompson's favorite example was of the jellyfish, which has a shape just like that of a blob of water falling through a column of oil. As the heavier substance descends, it takes on a specific flowerlike shape, just as the jellyfish does. The Canadian-born biologist Brian Goodwin agrees. His favorite example is of phyllotaxis, the kind of spiral-like patterns you see in the seeds or flowers of many plants, like the sunflower or the pinecone. Goodwin points out that these patterns do nothing for adaptive advantage and are instead simply mathematical patterns—technically, lattices—that are produced by the mechanical production of the plant. Order—but no real biological significance.

I guess in a way I am picking up on the division that Martin was talking about—that between form and function. People like D'Arcy Thompson—and by the way, it is hardly any surprise that Gould was keen on his thinking—are very much into form, structure. For them, the similarities between organisms come straight out of obedience to the same laws of nature. They—we, because I want to include myself here—think that that is the key to understanding life. We don't want to deny function or natural selection, but we do think it overrated by Darwinians. Let me say also that I think that physics and chemistry might be even more important than all of this. I realize that today the science-religion relationship focuses almost exclusively on the life sciences. Once, it used to be the case that the physical sciences were at the front of the battle, but no one today really doubts the Copernican revolution. Or associated sciences, like geology and plate tectonics, for instance.

Rudge: Surely, that is not quite true. Although the Reverend Hal might agree to the big bang in some sense, he would want to collapse the time scale down from fifteen billion years to a mere six thousand, and I suspect that he would want to modify the claims of plate tectonics also. He does not think that the earth's surface is made up of megasheets that slowly slide around the surface. At least, he does not think that they slide at the slow rate accepted by most geologists, and, at some point, he thinks there was a universal flood rejected by geologists.

Wallace: Well, that is all true, although I would say that biology is my main concern. I agree that I think that a lot of modern physics and geology is wrong, but it is evolution that really worries me.

Fentiman: Can we stay with Emily for just a moment? Suppose we do buy into what you have been saying. What difference does any of this have for the science-religion relationship? In particular, why or how does an integrationist like you—someone who thinks that ultimately science and religion are one—get turned on by order for free? What's the big plus over natural selection?

Matthews: Glad you asked! Obviously at one level, the big plus is that we can now get a level of intricate organization that I simply don't think is possible with natural selection. I'm not the only person who worries about this. We've mentioned the evo-devo people. They, too, think that natural selection is going to have to take a second-place position in the light of new scientific discoveries. Remember the point about human DNA and fruit-fly DNA being virtually identical. This has led three of the leading researchers to say the following:

> The homologies of process within morphogenetic fields provide some of the best evidence for evolution—just as skeletal and organ homologies did earlier. Thus, the evidence for evolution is better than ever. The role of natural selection in evolution, however, is seen to play less an important role. It is merely a filter for unsuccessful morphologies generated by development. Population genetics is destined to change if it is not to become as irrelevant to evolution as Newtonian mechanics is to contemporary physics.

But you cannot just leave things as that. Agree that organisms are built on the Lego principle, with the same parts being used over and over again. You have still got to have the organization imposed on all of this—the organization that gives you as the finished product the White House or a spaceship. Hal here wants to say that this is the work of the Intelligent Designer. I agree with the scientist and the philosopher on our panel that that just isn't on in modern science. So, since natural selection can't do the job, I turn to order for free and physics and chemistry.

But remember that I am an integrationist, so I look for theological, as well as scientific reasons, for my choices. For me, ultimately, they are all part and parcel of the same thing. Theologically speaking, I think it is generally agreed that the biggest problem that people have with the idea of God—especially with the idea of the Christian God—is the problem of evil. Why does a good and powerful creator allow evil? Now, we Christians do have some answers to this. Most particularly, moral evil—the evil of the Nazis that led to Auschwitz—is the result of original sin. Hitler and his dreadful crew had the choice between good and evil, and they chose evil. It broke God's heart, but that was the result of his loving decision to make us humans free beings. We are not just robots.

Davies: Hang on a minute. I can't let you get away with this. In his new book, *The God Delusion*, Richard Dawkins establishes beyond doubt that Hitler was a convinced Roman Catholic. It was Christianity that led us to Auschwitz—to the Holocaust. It's not just a question of our being free beings, but of our being free *Christian* beings. Or rather of you people being *Christian* beings. If it hadn't been for Christianity, we wouldn't have had these appalling evils in the first place. That's why we need a world run by science. It may be a tough world, one with no easy solutions. But it's a decent world.

Matthews: Now, I'm really mad. I've been putting up with these jibes for three programs now. I'm sick of it! Dawkins's book is not just wrong or biased. It's obscene. Hitler was born a Catholic, but nothing in National Socialism follows from Catholicism or any other branch of the Christian religion. Hitler himself was contemptuous of religion. He said so again and again. It is for weak people. And if not Hitler, what about Stalin and Mao? Even greater monsters and the causes of untold misery and death. They were certainly not Christian or followers of any other religion.

Of course there were individual Christians—too many individual Christians—who supported the Nazis. From Catholic prelates to Lutheran pastors. Let's not forget our own Father Coughlin, who used to broadcast vile pro-Hitler, anti-Semitic, radio sermons during the Depression. But, there were also Christian saints, Christian martyrs during the Nazi reign. Pastor Dietrich Bonhöffer chose to return to Germany to witness against Hitler. He ended his life at the end of a rope, just before the war ended. Or what about the White Rose group, that small collection of Christians in Munich in 1942 and 1943 who distributed pamphlets against the Nazis? Hans Scholl and his sister Sophie died on the guillotine for their actions and their faith.

You scientific atheists, you make me sick. I don't care about you being scientists. I don't care about you being atheists. I do care about you claiming to be purer than pure. You distort and distort. You are not worthy to lick the boots of those people. You're hypocrites. You're what Jesus called "Whited Sepulcres."

Wallace: I am glad that we have seen in recent federal elections that most people are opting for moral values rather than the godless atheism of people like Professor Davies.

Matthews: I am not much keener on you, frankly. If you would stop obsessing about gays. Start feeling ashamed of the fact that your leading college for ministers—the Southwestern Baptist Theological Seminary—fired its only female theology professor because she was teaching men. Worry about sick kids and not about ten-day-old fetuses. Then, perhaps the rest of America would start taking you seriously when you talk about right and wrong. Haven't you ever heard of the Sermon on the Mount?

Wallace: And if you were a Bible Christian, you would know that there are things that are absolutely prohibited. Saint Paul said explicitly, "Let a woman learn in silence with all submissiveness. I permit no woman to teach or have authority over men; she is to keep silent."

Matthews: Yes, and he also said in his Epistle to the Galatians: "There is neither Jew nor Greek, slave nor free, male nor female, for you are all one in Christ Jesus."

Fentiman: Time to pour some oil on troubled waters. We want debate here, not a shouting match. No, Professor Davies, I'm not going to let you get in now. We have an agenda, and I want to get back to it. We're talking about science and religion, not just religion. So, let's cut back on the Bible quoting. Emily, you were talking about the problem of evil. You've spoken about moral evil. I seem to remember from my philosophy student days that the other kind of evil is physical evil. What about it?

Matthews: Physical evil has always been more of a problem. Why does God let earthquakes kill so many people or tsunamis drown innocents? Actually, as a Whiteheadian, I have some answers, because my God is not the all-powerful God of the philosophers, but it is still a problem. And many people think that evolution—natural selection, particularly—exacerbates this problem. Darwin thought so, and Richard Dawkins thinks so. Dawkins thought that Darwin made it possible to be an intellectually fulfilled atheist, and Dawkins thinks that selection makes it obligatory. He is always quoting a letter that Darwin wrote, around the time of the *Origin*, to his American friend Asa Gray, professor of botany at Harvard and a committed evangelical Presbyterian. I have brought it with me to quote:

> With respect to the theological view of the question; this is always painful to me. I am bewildered. I had no intention to write atheistically. But I own that I cannot see, as plainly as others do, and as I should wish to do, evidence of design and beneficence on all sides of us. There seems to me too much misery in the world. I cannot persuade myself that a beneficent and omnipotent God would have designedly created the Ichneumonidae with the express intention of their feeding within the living bodies of caterpillars, or that a cat should play with mice. Not believing this, I see no necessity in the belief that the eye was expressly designed.

Actually, Darwin then goes on to say that he still believes in some kind of god, but no matter. The damage is done. Natural selection seems to lead to pain and suffering. At the worst it is evil. I believe the evolutionist George Williams refers to nature as a "wicked old witch." At best, as Dawkins says, nature is indifferent. Neither good nor bad, but meaningless. "If Nature were kind, she would at least make the minor concession of anesthetizing caterpillars before they are eaten alive from within. But, Nature is neither kind nor unkind. She is neither against suffering nor for it. Nature is not interested one way or the other in suffering, unless it affects the survival of DNA . . . . The total amount of suffering per year in the natural world as beyond all decent contemplation." And then he goes on to say: "As that unhappy poet A. E. Housman put it: 'For Nature, heartless, witless Nature; Will neither know nor care.'" DNA neither knows nor cares. DNA just is. And we dance to its music.

It seems to me that if self-organization is the main way in which we get evolutionary change, then things are altered significantly. Evolution can be seen to be a much more creative and positive process. It is an unfurling rather than a hating, bloody struggle for existence. God has incorporated the seeds and forms of life in the laws of physics, and now all that is needed is time to bring them to fulfillment. As the Regius Professor of Religion at Oxford University, Keith Ward, has said, this is a kinder, more gentle form of evolution than harsh Darwinism, and thus much more in tune with Christian thinking.

Davies: Except, of course, it is not true!

Matthews: Well, that's what you say!

Davies: Emily. You're in the same position as Hal over there. You want to criticize him for twisting science to his own ends, cherry-picking the bits that he likes, but you are doing exactly the same thing. Self-organization is at best a minor phenomenon and, in any case, does not deny the pain and suffering in the world. The cheetah chasing the antelope, for instance.

Rudge: I want to add that I think Emily and her chums are being a bit unfair to Darwinism. Although Darwinians stress adaptation, they never claim that everything is perfectly adaptive.

Matthews: I never said that everything is perfectly adapted. Indeed, my theology pulls back from that—the thinking that an all-powerful god will make everything perfect, straight off. The whole point of Process Philosophy is that life is a state of becoming rather than being.

Rudge: I stand corrected. But, it is just as well that you made that qualification. With evolution, you do not expect to see perfection in design all of the time. Especially with natural selection. The old joke is about the bear chasing you in the woods. You do not have to be absolutely fast. Just faster than the chap next to you! There are lots of organic parts that are good enough, but not perfect. The male urogenital system, for instance. The sperm duct is hooked around the urethra, like a garden hose hooked around an apple tree, rather than going straight from testicles to penis. It works, but as any old man will tell you, it is not a perfect system. This really is the big theological problem. If you have God doing the designing for the irreducibly complex, why didn't he clean up some of the simple problems while he was at it? Why didn't he cure some of the horrendous genetic diseases of humankind—like Tay-Sachs disease—that kill little kids, especially when it only needed one or two molecules to be shifted around?

Wallace: God works in mysterious ways, his wonders to perform.

Rudge: A bit too mysterious for folk like me, I am afraid. Here, I do have some sympathy for Emily. As soon as you bring God down into the ongoing-creation business, claiming that he can do anything, you run into theological problems. If you keep God distant in a way, he created and then let things run their course, you are better off. Or if you adopt Process Philosophy, and make God a limited worker with humans in the creative process, you might be able to get away with it. It's not God's fault that there is Tay-Sachs disease. He is working with us to find a cure.

Wallace: And if you really think that this is the Christian God, you are as mistaken as the Reverend Emily Matthews!

Fentiman: Can I pull the discussion back, please? David, you have been rather contemptuous of self-organization. Could you talk a little more about that? Emily likes it because she thinks it helps to explain the intricate designlike nature of organisms and to get around the problem of evil. Yet, you as a scientist want to argue that it is insignificant and so her gambit won't work. Am I right?

Davies: Right on! I think that Emily and her pals have got a bad case of what Haldane used to call "Aunt Jobisca's theorem": it's a fact the whole world knows. In her case: it's a fact the whole world knows that natural selection cannot create intricate design, complex organization. Well, all I can say is: News to me and to a lot of other evolutionists! Richard Dawkins is always saying, correctly, that the biggest mistake is to assume that because you cannot see how natural selection might have done something, it does not follow that natural selection could not have done it.

Wallace: I am not sure I would use Dawkins as my reference here. Like Emily, I don't see how a mechanism like natural selection can produce real design—or even seeming design. The kind of functioning we were talking about at the end of the last program. Go back to Dawkins's computer program that produced METHINKS IT IS LIKE A WEASEL. You have agreed that it is not really a fair analogy. And it couldn't be. You say you have random mutation and a blind process like natural selection, and yet you turn out the eye! Nonsense!

Rudge: Nonsense, perhaps; true, certainly. When someone like Behe says that blind law cannot create designlike complexity, he is just wrong. Leave Dawkins's claim about lines from Shakespeare, although I do think understood in a limited sense, it does have power. It points out that you can generate order in a reasonable time if you have a selecting and keeping or recording mechanism. More generally and excitingly, people who are working on artificial life have made real strides in the right direction. For instance, a theoretical biologist called Thomas Ray set up a computer program with simple forms subject to a kind of random mutation force and their being selected if they found some way to propagate more efficiently than others. Before very long, the forms developed all sorts of sophisticated mechanisms, like being able to repair themselves and sitting parasitically on the successes of others. An artificial model admittedly, but complexity from simplicity without an intelligence building it in.

Fentiman: Can we get back to the main point here? Professor Davies, you think that natural selection can produce really complex organisms. Is there nothing then to the ideas of Gould and Kauffman and Goodwin? Are physics and chemistry entirely irrelevant?

Davies: It's not a question of whether physics and chemistry are irrelevant, but whether they are all-powerful and all-sufficient. All biologists agree that the physical world sets what they call "constraints" on the nature of organisms. Organisms cannot be purely adaptive all of the way and all of the time, because physics and chemistry just won't let this happen. A nice example comes from the comparison between cats and elephants. Why do you never see a cat as big as an elephant? The answer is simple. As the height goes up in a linear fashion—one foot, two feet, three feet—the volume, and hence the weight, goes up in a cubed fashion—one pound, four pounds, nine pounds. That is just a matter of simple physics. So, if you get as big as an elephant, you need elephant-sized legs to hold you up. The slender jumping legs of the cat just will not do. They would break given the weight. So, cats are constrained by physics to be relatively small. Note how the big exceptions to this are the whales that get superbig. They live in water and so Archimedes' principle keeps their comparative weight right down.

Now let's push the point a bit. Quite frankly, as people often said about Gould, why should any of this be news or threatening to Darwinians? They have known all along that adaptation is never the best possible. Remember, as we were just told, it is always a matter of being better than the chap next to you! The real challenge comes from the kind of thinking that so excites Emily—the self-organization people like Stuart Kauffman and Brian Goodwin. Note that there is often some ambiguity in their thinking. Some of the self-organization types rather suggest that adaptation is irrelevant. Organisms have the shapes that they have because of the laws of physics and chemistry—the jellyfish, remember—and there is nothing here about helping the organism. Such fish are as they are, and they can like it or lump it. However, I don't think most self-organization supporters are inclined this way. I suspect that they think that the shapes are adaptive, but that physics and chemistry do all that is necessary and natural selection has no role to play.

The Darwinian response is simply that a lot of shape—or form, as it is often called—is indeed due to physics and chemistry, but it only survives because natural selection lets it. If the jellyfish's shape did not do what was needed, then jellyfish would be extinct. In truth, more subtly, what we generally have is a shape provided by physics that is then molded by natural selection to the organism's ends. Phyllotaxis, the patterns shown by the sunflower, is a case in point. Everyone agrees that the patterns follow very strict laws—somewhat curiously, they key sequence determining different patterns is the Fibonacci series, where the next number is the sum of the previous two numbers. So we have 0, 1, 1, 2, 3, 5, 8, 13, and so forth. This formula was worked out by a twelfth-century Italian mathematician, Leonardo Fibonacci, who wanted to know the offspring number in each generation of an initial pair of breeding rabbits. Of course, today it is better known because it was one of the vital clues in Dan Brown's best seller, *The Da Vinci Code*.

What Darwinians say is not that this is not right—plants do show the Fibonacci series and this is a mathematical function of the way they grow. Nevertheless, selection still works on the product. By altering the tilt of the leaves to catch the sun and the size of the seeds to aid in distribution and so forth, adaptation comes flooding right back in. Order is never free. Ultimately, we Darwinians think people like Kauffman spend too much time in front of the computer screen playing with algorithms and not enough time out looking at real organisms. Either way, there is no solution to the question of organization here. As pertinently for someone like Emily, there is no solution here to the problem of evil.

Rudge: I'd like to jump in. I agree with Emily that the problem of evil is a massive problem. But, I also agree with Dave that self-organization is not going to solve it. At best, it nibbles at the edges, and at worst, is nonexistent.

My personal inclination, as one who keeps things separate, is not to go to the evil question at all. Nature is as it is, and religion is as it is. If a religion like Christianity cannot find a solution to evil on its own terms, biology is neither going to help or hinder. To be fair, a religion that has the suffering of God right at its center is hardly a religion indifferent to the problem of evil. Whether it can solve it or not is another matter.

But this I would say. If you do seek some kind of philosophical or theological solution to the problem of evil, I am inclined to think that natural selection might help rather than hinder. The most popular solution is that associated with the great German philosopher Leibniz. He argued that although God is all-powerful, this does not mean that he can do the impossible. He cannot make two plus two equal five. And when God created, the impossibility factor raises its ugly head there, too. If God wanted to make fire burn and be useful to us, then it was a good thing to attach pain to the feeling of burning. Imagine if we could sit on a hot stove without feeling anything! It would be devastating for our health. Likewise, if fire is going to be able to burn wood and coal, for our benefit, it also is going to burn houses and forests for our discomfort. It is a package deal. God is trying to maximize the good things, but for him, as for us, it is a matter of taking a bit off in one place to get more in another place.

Davies: I thought Voltaire put the boot into that line of thought. Wasn't that the whole point of his parody *Candide*? Dr. Pangloss, Steve Gould's favorite, was always saying when things went wrong that it was for the best of all possible reasons in the best of all possible worlds. Gould, as we know, was using it to parody those Darwinians who say that everything must have an adaptive function—it is the best of all possible designs in the best of all possible worlds.

Rudge: I am not a Christian, but speaking on their behalf, I guess that they would say that this is an exaggeration, just like Gould's characterization of Darwinism was an exaggeration. But, suppose for a moment that the Leibniz argument does work. Then, it is plausible to suggest that natural selection might have been the only mechanism that would produce designlike features on a wholesale basis. This at least is the claim of Richard Dawkins of all people: "If a life-form displays adaptive complexity, it must possess an evolutionary mechanism capable of generating adaptive complexity. However, diverse evolutionary mechanisms may be, if there is no other generalization that can be made about life all around the Universe, I am betting that it will always be recognizable as Darwinian life." In other words, according to

Dawkins, if God did create through law rather than in one step by miracle—
and I am sure that Emily can think of good theological reasons why this
might have been so—then, it had to be through natural selection, and this
brings with it the problems of pain and so forth. Admittedly, it might not explain all of the physical pain, but it goes a long way to explain it in the biological world, which is what we are looking at right now. Far from Darwinism intensifying the problem of evil, perhaps it relieves it a bit!

Fentiman: Very nice! One might almost think you had training by the Je-
suits! I'm now going to exercise my power as the moderator and move us on
again. Thus far, we have been focusing on biology. And, we will next time
again, as we move toward discussion of humans. Although I suppose we also
might have something to say about the social sciences. For the rest of this
program, let's switch to the physical sciences. We all know that it was the
physical science side that caused the big science-religion upheavals four hundred years ago. Can the earth really be moving? Is it no longer the center of
the universe? Did the sun really stop for Joshua, and if it did, doesn't this
mean that it is usually moving? I guess most people are past all of these now.
But, there are still conflicts. Obviously, as was brought up last time, someone
like the Reverend Hal is not satisfied with the way physicists and geologists
and others date the major events in earth history. Are there other issues that
come up? Are there any new issues that physics, for example, raises that make
us think of implications for religious belief?

Matthews: What about the anthropic principle? It seems to me that this
surely shows that the dismissive attitudes of Professor Davies are misplaced.
In fact, I am inclined to think that this principle should make us go back and
look again at the supposed expulsion of real design from biology.

Fentiman: We're getting ahead of ourselves a bit at the moment. Can someone tell me what we are talking about?

Wallace: Let me come in here. As you know, I trained as a physicist before I
was called to the ministry. Let's start with the basics. The so-called anthropic
principle, something which apparently was first named by a physicist called
Brandon Carter in 1973, comes in various forms. I will use the taxonomy introduced by John Barrow and Frank Tipler in their very useful book on the
subject. First, they talk of the weak anthropic principle, the WAP. They say
this of the WAP: "The observed values of all physical and cosmological quan-
tities are not equally probable, but they take on values restricted by the re-

quirement that there exist sites where carbon-based life can evolve and by the requirements that the Universe be old enough for it to have already done so." Basically, this does not say a huge amount and is perhaps almost tautological. It says that if you have life like we have, then the conditions for it must have been such that life like we have appears and is sustainable. Lots of people had spotted this, way before Carter named it. For instance, Alfred Russel Wallace, the codiscoverer of natural selection along with Darwin, wrote at the beginning of the twentieth century: "Such a vast and complex universe as that which we know exists around us, may have been absolutely required . . . in order to produce a world that should be precisely adapted in every detail for the orderly development of life culminating in man." L. J. Henderson, the Harvard biochemist early in the twentieth century, wrote a whole book called *The Fitness of the Environment*, meaning the fitness for life.

More challenging is the strong anthropic principle, SAP. Barrow and Tipler characterize this in a number of ways, but the key idea is that the universe had to be "fine-tuned," to get life going at all and sustain it. The various constants that govern the laws of nature could not be chosen at random, but had to be very exact within incredibly narrow limits. "There exists one possible Universe 'designed' with the goal of generating and sustaining 'observers.'" In other words, the lack of randomness implies a designer of some sort. What constants are we thinking of? Well, gravity for a start. It is 10^{39} times weaker than electromagnetism. That's why when you hold up a rock, the hammer in your other hand doesn't immediately jump over to the rock! More seriously, if gravity had only been 10^{33} times weaker than electromagnetism, the suns of the universe would be a billion times less big and burn a million times faster. The nuclear weak force is 10^{28} times weaker than gravity. If it had been slightly weaker, the hydrogen of the universe would have been converted to helium, and that would have meant no water. Hardly propitious for life as we know it!

The list continues, but the point is made.

Fentiman: Do I understand that you accept the anthropic principle, perhaps in some strong form?

Wallace: Oh, certainly! Although, of course you will realize that by now this is all part of my overall world picture. I certainly see physics here supporting my religion, which is what you would expect from my dialogue position—my mutually supportive position—on science and religion.

Fentiman: Time now to let the other side have their say. Professor Davies?

Davies: I hardly know where to begin. Some people have complained that the SAP is not testable, not falsifiable, and, hence, not science. I agree, but from my perspective this is not a devastating criticism. I don't think it could be science! If it were, science could prove God and my conflict position would be destroyed. I think the argument is metaphysical and, as such, should be fought at that level, using science as needed. For a start, note that if the argument does work—if there is an Intelligent Designer—then, this triggers all of the kinds of criticisms leveled by people like the eighteenth-century philosopher David Hume. He went after the design argument as he knew it in his day, and he is just as pertinent today as back then. He raised questions that affect the anthropic principle, as well as the ID argument in the biological world. Do we have one designer or a squad of designers? Is there a trail of botched attempts and are we just one attempt on a course to a perfect universe? What about the problem of evil? Is this something that had to be? And so forth.

But, I suspect that there are deeper problems with the anthropic principle in its strong form. At least, some eminent physicists tell me this, and I am inclined to agree. I should say that I am suspicious of the whole business. There is too much of the argument from ignorance around the whole thing to make me easy. I've got the Richard Dawkins worry once again: "Nature is cleverer than we think." I suspect it applies here. Does life have to be carbon based? Have we an ontological proof for this claim?

Fentiman: An ontological proof?

Rudge: It means a proof about existence. The most famous ontological proof is for God's existence, something that tries to derive his existence from his definition. As Descartes said, if God is perfect, then he must exist because existence is a perfection. Kant challenged this by arguing that existence is not a property like red or round, and so existence is neither perfect nor imperfect.

Fentiman: Thanks. I feel a lot wiser now! Keep going, Professor Davies.

Davies: I am trying to show that strange or seemingly designed physical constants don't necessarily call for intelligence. Take an analogy from the biological world. Cicadas have gestation periods of either thirteen or seventeen years. How do they choose these prime numbers? There must be a thinker behind them, just as the heroine of the movie *Contact* says there must be a thinker behind the prime numbers she gets from outer space. But, for the cicadas there isn't! It turns out that selection favors these primes because then it is difficult for the predators to get into sync. If the period were twelve, for instance, then even if the predators had, say, only a four-year gestation period, they could score big time every third time. Prime numbers make this nearly impossible, so is a good adaptive strategy for the cicadas. Analogously, I wonder if there are like principles making our universe. Looks designed, but isn't really.

Steven Weinberg, the Nobel physicist whom we've heard about before, points out that if the SAP really works, then we must be sure that the constants are "fine-tuned," and that there are no other possible universes. If there are the latter, then maybe ours is one of many, and while it is true that it bears life, the fact that it does may be no more than chance. And it is no argument to say that it is designed that we are on it. That is like looking for design by a lottery winner. Somebody had to win, but there was no design that it was Joe Bloggs who was the winner. It may be chance that it is our universe, but it is not designed that there is such a universe, and, after all, if we are here then by the WAP, we had to have the conditions of life. As far as the fine-tuning is concerned, Weinberg argues that often things are not as precise, nor need they be as precise, as people think.

For instance, the formation of carbon is often cited as a case where super-accuracy was needed. To make carbon from helium, you need a huge energy state above normal—in fact, about 7 million electron volts (MeV) above normal. However, it also turns out that if you go over 7.7 MeV, things won't work properly. Miraculously, apparently, it turns out that there is such a needed state for carbon that comes in at 7.65 MeV. Now, think about it. Carbon misses the cutoff by 0.05 MeV, or less than 1 percent. Surely, this cannot be chance. Over the cutoff and no life. Under the cutoff and abundant life. But hang on a moment, says Weinberg. Why the figure of 7.65? It turns out that this is a function of the carbon production—first, you combine two helium nuclei to make beryllium, and then you bring in a third to make carbon. And here, apparently, there is more flexibility. The beryllium-helium join up occurs at 7.4 MeV. So, you cannot go more than 0.3 MeV more before things come apart. But, this means that in fact, the carbon state (at 7.65) misses the target by 0.05 MeV. But, remember this is against a rise of .25 MeV (from 7.4 to 7.65). This means that the upper target of 7.7 is actually missed by 20 percent (0.05/.25) rather than 1 percent. So in other words, while it is true that the 7.7 figure is fixed as we know it, further down the road things are nothing like as tight and the coincidences seem less striking.

What about the uniqueness factor? A lot of people today are drawn to the idea of multiverses. Perhaps there are indeed lots of different universes, and so ours is not so surprising. It is odd that it does sustain life, but it is no more odd than that, between 0 and 100, there is only one number that can be divided by both 9 and 11. Let me read something that Weinberg wrote in the *New York Review of Books*:

> Recent developments in cosmology offer the possibility of an explanation of why the measured values of the cosmological constant and other physical constants are favorable for the appearance of intelligent life. According to the "chaotic inflation" theories of André Linde and others, the expanding cloud of billions of galaxies that we call the big bang may be just one fragment of a much larger universe in which big bangs go off all the time, each one with different values for the fundamental constants.

In any such picture, in which the universe contains many parts with different values for what we call the constants of nature, there would be no difficulty in understanding why these constants take values favorable to intelligent life. There would be a vast number of big bangs in which the constants of nature take values unfavorable for life and many fewer where life is possible. You don't have to invoke a benevolent designer to explain why we are in one of the parts of the universe where life is possible: in all the other parts of the universe, there is no one to raise the question.

Rudge: Are you claiming that these theories are all true and that the strong anthropic principle is, therefore, false?

Davies: Probably not quite as strong as that. I am saying that they are at least plausible and that, hence, one should not take the SAP as definitive. I would also say that given the other evidence, the failure to find design in the living world and so forth, the presumption is against God and so that should be the null hypothesis here.

Matthews: I am not sure I agree with you. I know that Weinberg has said that the more he studies the universe, the more interesting he finds it and the less reason to think that there is some ultimate meaning—or I guess one should use a capital M, and say Meaning. I just disagree here. I think the final question— Why is there something rather than nothing?—is still there.

Rudge: In a sense, I am inclined to agree with you, Emily. I am with David against reading too much into the anthropic principle, however construed. But, I don't think that science even asks the ultimate metaphysical questions. This goes along with my whole feeling about the nature of science. I was going to explain it at the beginning of the first program, but we ran out of time.

Fentiman: I am afraid we are starting to run out of time here, too, so your general views on science are going to have to wait. But I promise you that we will get to them before we finish. For now, getting back on track, I take it that you are now finished, David.

Davies: I have not yet talked about the completely ridiculous anthropic principle, spelled—

Fentiman: I think we can leave that to the imagination of our audience. This is public television!

Davies: Oh, I thought I was allowed four-letter Anglo-Saxon words!

Rudge: Interestingly, it is not Anglo-Saxon, but it first came into use at the end of the fifteenth century in Shropshire, a county in the English Midlands, and meant the dregs from a barrel of ale. It is a corruption of the Latin *crappa*, meaning "chaff," as one gets after the threshing of the harvest. It went out of use in England in the seventeenth century, but went to America with the pilgrims, where apparently it took on its present meaning. Now, the other word that is synonymous is in fact Anglo-Saxon, and goes back before the Normans, when there was the verb *scitan*, coming from the Indo-European root *skei*, meaning "to cut" or "split." Apparently "science" also comes from that.

Wallace: It seems to me that there are times when you professors have altogether too much knowledge.

Matthews: It is not a question of too much knowledge or even of professors. It is more a matter of guys trying to show off!

Rudge: Ouch!

Fentiman: Okay, children all around. We are just about out of time today. Next week, we will take up issues in the history of life, leading into our discussions about humans.

Wallace: Ah, we are going to discuss the Completely Ridiculous Understanding of Development, pronounced "CRUD."

Davies: Very nice, and before Martin Rudge gives us another lecture on etymology, I am going to say farewell on everyone's behalf. Until next week, this is Redvers Fentiman speaking for David Davies, Martin Rudge, Emily Matthews, and Harold Wallace. Good night!

CHAPTER FOUR

~

Program Four: Histories

Fentiman: Good evening, and welcome to the fourth program in this series of *Eternal Questions*. We have been discussing the relationship between science and religion, and tonight we are going to turn to life's history, concluding with a look at the most important issue of them all, namely, us! My name is Redvers Fentiman, and my panel is the same as for the first three programs.

Welcome, everyone, and welcome to our studio audience and to you at home. Let's get down to business. We are going to focus tonight on the history of life. I am going to assume that we all accept the basic picture, sketched for us a couple of programs back by Professor David Davies, our resident evolutionist from the Massachusetts State Institute of Science. At least, we all accept that this is the basic picture of evolutionists. There were rather primitive aquatic organisms, and then they invaded the land. In succession, we see the amphibians, and then the reptiles, and finally the mammals. Humans seem to have come last and to be very recent. I don't want to get into controversy, but I am sure we are all agreed that the dinosaurs were reptiles, and I believe that common opinion today is that the birds came from the dinosaurs. At least, that is the case if you believe in evolution. And I suppose a similar tale can be told about plants, with the flowering plants coming last.

Now, I am going to turn first to the Reverend Hal Wallace, pastor of Rollingbrooke Stones Baptist Church in Atlanta. Even before we get to the evolution of humans, my understanding is that evangelicals like you, Hal, think that there are major problems for the evolutionist. I believe that a couple of programs back you mentioned the early history of life and then the

Cambrian explosion. Would you elaborate on these, please, making sure to introduce the topics so our audience knows fully what you are talking about?

Wallace: Absolutely. You evolutionists have claimed that life began on earth within a billion years of its formation, that is, about 3¾ billion years ago. Now what happened after that? Everybody agrees that there was a huge increase in life forms at the beginning of the geological period known as the Cambrian. There was such an increase that it was known as the "Cambrian explosion." You evolutionists think that the explosion was rather more than five hundred million years ago. I don't, of course, but let's leave that for the moment. My point now is that, on your terms, there was a period of over three billion years when life flourished—existed, at least—before the explosion. My question is: Where is the evidence of all of this? Where are the fossils now? Your hero Charles Darwin admitted that there are none, and he proposed a desperate, ad hoc, face-saving device. He said that the animals and plants would have lived in the areas where there are now oceans, and even if we could look at the submerged rocks, the pressures would have destroyed all traces of life. Professor Davies, you claim to be a Popperian. If ever there was an unfalsifiable hypothesis, it is this!

Things are worse than this, though. What about the Cambrian explosion? At one point, there was no life. Moments later, there were all kinds of life forms. And not simple ones, either. Very complex creatures, like trilobites. Anyone who says that these were simple has obviously never looked at one. All of those segments right down the body. And their eyes! Really sophisticated compound lenses, of a design that was worked out by the French philosopher René Descartes—a design that made the creatures able to focus the light into a sharp point rather than being diffused because of the different refractive indices of the various components of white light. Or what about the book written by Stephen Jay Gould, *Wonderful Life?* It is all about the Burgess Shale in British Columbia, Canada. This deposit contains huge numbers of soft-bodied animals, each and every one of which is very complex and highly adapted to its habitat. Gould's book was written about twenty years ago, and, in the time since then the story, if anything, has become more detailed rather than less.

You see, you scientists and philosophers have been sneering at me over my claim that I alone am taking seriously the need to harmonize science and religion and use them to each other's end. No, no! Don't try to deny it. I don't take it particularly personally. But in truth, I am the only one who is looking seriously at the scientific facts and trying to draw the inferences. Because I am not hamstrung by the dogmatic assumption of naturalism—

methodological or metaphysical—I can think outside the box, as it were. I can see that any scientific theory simply cannot address these issues, certainly not any evolution-based, scientific theory. We've already been over these issues at one level. Professors Davies and Rudge think that natural selection can do everything. Reverend Emily Matthews thinks that we need some kind of physics-based mechanism.

I am not even going to criticize either of these positions since you have done such good jobs yourselves! What I do want to say is that the Cambrian explosion intensifies these worries and problems. At one point, there was nothing. Then, there was not just something, but almost everything. In other words, whatever it was that caused things to appear was not some slow, leisurely process. It's improbable that it was either natural selection or the workings of physics and chemistry. So I'm free to go with the only sensible solution, namely, that there was divine intervention. And this, of course, harmonizes perfectly with what I learn from the Bible.

We've already had the J. B. S. Haldane crack about not finding mammals way down in the fossil record. I know that you proudly say that because of evolution you don't expect to find humans down in the Cambrian. As a general rule, neither do I! You must remember, as far as I am concerned, the real cause of the fossil record was the Deluge. As Whitcomb and Morris say in *Genesis Flood*, what you find is that the more primitive organisms got caught first at the bottom of mountains, and then in succession as they went up to the peaks, more and more got trapped by the rising waters. I expect to find humans at the top of the fossil record, because they were the ones who stood at the summits and drowned last of all. If, like everyone else, I might be allowed to quote:

> It is reasonable . . . in the light of the Flood record, to expect that vertebrates would be found higher in the geologic column than the first invertebrates. Vertebrates in general possess much greater mobility, and this factor, together with their pelagic habits, would normally prevent their being entrapped and deposited in the deepest sediments.

So you must understand, at one level, I don't at all want to challenge the story of life that our moderator just gave us at the beginning of the show. I have told you all along, I am with science—with good science—not against it. I agree that the amphibians came before the reptiles, came before the mammals, came before the humans. It's just that I don't give this record an evolutionary interpretation. I want to add that there is a potential crucial test between my position and that of someone like Professor Davies. I expect sometimes to find anomalies in the record. Perhaps some human was a little slow going up the mountain and so got stuck back with the dinosaurs. Professor Davies would deny that this could ever happen. According to him, the dinosaurs went extinct over sixty million years before the first humans. And, when we turn to the record, what do we find? Clear evidence of humans and dinosaurs coexisting! In the Paluxy riverbed, near Glen Rose, Texas, there are human footprints and dinosaur footprints together! Score one for me and my side, and take one off Professor Davies and his side. Although, truly, take everything off Professor Davies and his side. According to him, this simply could not happen. I should say incidentally that the Paluxy River Bed does more than refute the evolutionists. It offers positive evidence for what I believe. Some of the human footprints are absolutely huge. Why am I not surprised? Because the Bible tells us, in Genesis 6:4, that there were "giants in the earth in those days."

I know that Professors Davies and Rudge will never give up, and I suspect that the same is true of the Reverend Emily Matthews. But, I beg those of you in the audience whose minds and eyes are not firmly shut, to look at the facts and see what they tell you.

Davies: And, I beg those of you in the audience whose minds and eyes are not firmly shut, to make sure you are not looking at the facts through the pebble glasses of religious fundamentalism before you say what you can see. I have never heard so much nonsense in all of my life.

Fentiman: Don't just emote. Give us some arguments.

Rudge: Well, what about a quick comment on the topic of those footprints. The Intelligent Design people, with some few exceptions, have never wanted to get into that sort of stuff. We've heard that Michael Behe, the author of *Darwin's Black Box*, is a Roman Catholic and, as such, has never been a biblical literalist. He takes on board a large amount of evolution, and certainly believes in a very old earth. It is just that he wants occasional interventions by the designer. So, he would have no desire whatsoever to find human footprints down among the dinosaurs. So, let's be quite clear, even if we take away an Episcopalian like Emily, some of the most conservative Christians in America—some of today's most prominent anti-Darwinians—are not into Flood geology like Pastor Hal.

The other thing is that the footprints are as phony as a three-dollar bill. At least, they are phony considered as human footprints. There is pretty strong evidence that the prize-winning cases—those that are given pride of place in *Genesis Flood*—were carved by locals during the Depression. The other cases are much more readily interpreted as partial prints left by the dinosaurs. This fits with reptilian gait rather than human-walking patterns, as well as the fact that if they were human prints, you would have embarrassing features to explain away, like big toes on the wrong side and that sort of thing. Indeed, even the Creationists are starting to cover their tracks, if one might use an appropriate metaphor.

Matthews: Do spare us the humor of philosophers. I feel like that eighteenth-century chap who said that he had often tried to do philosophy, but cheeriness kept breaking through.

Rudge: Sorry! Anyway, the point is that even the people at Henry Morris's organization, the Institute for Creation Science, have backed off and now only say that the human footprints are "possible" evidence of humans down with the dinosaurs. They are certainly covering something, even if it is only their backsides.

Fentiman: I think I need to intervene here before Martin gets a bit too anatomical. Why don't we swing back to David Davies and let him pick up the scientific gauntlet that Hal Wallace has thrown down.

Davies: Right. I'm not even going to get into the nonsense about the effects of the Flood. There just is no evidence of such an event. It may indeed be true that there were limited deluges. Perhaps the Black Sea was the result of a breakthrough in the straits by Istanbul, where the Black Sea today is connected to the Mediterranean. But, this is no worldwide covering of water, with some old chap with a long beard floating on the top. And, even if it were, the story about the animals climbing up the mountains is sheer nonsense. Apart from anything else, whether or not the dinos were hot blooded, we now know that many of them were a great deal more agile than your standard mammal, including your basic human. If it came to a race to the top, I would pit a *Tyrannosaurus rex* against a human any day of the week. I want to turn the Reverend Hal's sneer of a week or two ago back on him. This isn't science. It's Hans *Christian* Andersen.

Let's pick up the big questions, starting with the early fossil record. What was said about Darwin is true. He did not have pre-Cambrian fossil evidence and he did come up with some unfalsifiable, ad hoc hypotheses. I think he was a genius, but I don't think he was always right. But, let's not forget that he was writing 150 years ago and time has moved on since then.

As I see it, there are at least three lines of evidence which are, or which could be, relevant here.

First, there is the fossil evidence. Note that the fossil evidence is always a bit of a mixed thing for evolutionists. Duane Gish, the Creationist, wrote a book called *Evolution: The Fossils Say No!* Well, that is true, or rather, that is not true. The fossil record is an important element in the proof of evolution, but note that it is not the only element in the proof of evolution, and some—probably including Charles Darwin—think it is not necessarily the most important. Things like the Galapagos finches and the similarities between embryos of different species are just as important, if not more. Remember the point we made in an earlier program about the Darwinian consilience, where the fossil record is crucial in making out the paths of evolution, what we professionals call "phylogenies." How would we know, for instance, that the birds were descended from the dinosaurs without some fossil evidence? Archaeopteryx, that fantastic fossil discovered in the slate quarries of Germany just after the *Origin* was published, is simply a dinosaur with feathers.

But, the fossil evidence is not the only thing in working out phylogenies. Comparative anatomy, including comparative embryology, has always been very important. Why, for instance, did people even before Darwin, even before evolution, want to put humans off with the great apes—chimps, gorillas, and orangutans—rather than with cows and horses, or whales? Simply because we look more like apes than we do horses and cows and whales. It is a question of overall similarity. Whales have bigger brains than any of us, but we don't put humans together with whales, because in other respects we are much more apelike than whalelike. Embryology is important because of the point I just made, about the similarities at that level. Often, you can see similarities more readily in embryos than in adults—the chick embryo and the human embryo are alike—and so we can discern older relationships. We may not be very close to the chicks, but at some point in the past, we had shared ancestors.

The third line of evidence is that which comes from study of the external factors. The environment, for instance. Or the geology. Thanks to plate tectonics, we now know that the continents slip around the globe on large platforms, or plates. We mentioned this in talking about the origin of life. Do things like this slipping fit in with the fossil and comparative evidence and do they suggest paths or causes? The most obvious example is of the comet that hit the earth around sixty-five million years ago. We are certain that this happened and that it was down in the Gulf of Mexico. We know also that it would have disrupted the whole world, perhaps causing a kind of nuclear winter for a year or two until the dust settled. It is highly probable that this led to the end of the dinosaurs and the opening of the age of mammals. Incidentally, as I think we mentioned earlier in passing, mammals were around for a long time before that. Probably from about two hundred million years ago. It was just that when the giant reptiles were around, somewhat sensibly, they kept a low profile. Although, now we are starting to discover that from the first, they were a lot more inventive than you might expect. There were probably flying mammals—at least, gliding mammals—from a very early age. Indeed, the suggestion is that some of the really important branchings, creating the major groups of mammals, took place about a hundred million years ago, when the dinos were still very active.

Going back—or forward—to the end of the dinosaurs, it has to be admitted that after the comet hit, no one is quite sure what went on then. Did the dinosaurs starve, for instance? If they were not hot blooded, then in the cold they would have been virtually paralyzed and unable to forage, let alone copulate. Mammals, on the other hand, were small, nocturnal, and hairy. They could well have survived. A mouse needs a lot less food than Tyrannosaurus rex. One intriguing hypothesis comes from a fairly recent finding that reptile sex can be caused by the temperature at which the eggs develop—in some cases, males are produced by cooler eggs and females by hotter, and sometimes it is the other way around. Could it be that the cool temperatures after the comet hit the earth meant that all of the new dinosaurs were of one sex? Just like the Shakers, they died out because no one was having it off?

Anyway, enough preliminary.

Wallace: Preliminary? It seems to me that you are getting biblical in length, if not in conviction. Why don't you answer my questions? What about the pre-Cambrian fossil record? What caused the Cambrian explosion?

Davies: I am getting to that, and I am not going to apologize for doing the groundwork. Start with the pre-Cambrian. It is now simply not true that we have no fossil evidence at all. It is certainly true that we do not have huge amounts, but there is such evidence, and it goes back as far as 3.5 billion years ago—in other words, about a quarter of a billion years after the earth cooled to such an extent that life was even possible. From then, until the Cambrian explosion 3 billion years later, the record shows the kinds of upward changes that one would expect if evolution were true. I admit that you hardly see evidence of natural selection, but one sees the adaptive breakthroughs one would expect if a design-producing mechanism like selection were at work.

Mention has already been made of Lynn Margulis. Her brilliant hypothesis was that some primitive life forms joined together to make more complex life forms. This was way back when everything was just a question of organisms having one cell only. Long before we had many-celled, multicellular organisms. Margulis wasn't absolutely the first to think up the idea that sophisticated cells with many different parts with different functions—so-called eukaryotes—were produced by the combination of simpler cells, prokaryotes, but it was she who pushed the idea against opposition and disbelief in the modern scientific era. I should say that none of this is pie in the sky. Some cellular parts of eukaryotes are remarkably similar in genetic composition to some prokaryotes—too similar to be pure chance.

Matthews: Let me score one for my own sex. Lynn Margulis is a holist. She does not break things down into parts. She is not a reductionist. I talked a week or two back about the Gaia hypothesis, something that sees the whole world as bound up in one organic unity. I am not at all surprised that she is one of the greatest enthusiasts for this idea. I know it is even more fashionable to sneer at feminists than at Creationists, but great ideas and movements always start out in the minority. Who would have thought that a little black woman refusing to sit at the back of the bus would have sparked the civil rights movement? This is one of the major reasons why my God is as much a female as she is a male.

Rudge: Actually, I am not sure that you have to be a feminist to be in favor of a God who in some sense transcends sexuality. It was always one of the great distinguishing features of the Jewish Yahweh that he or she or It was not a sexual being like so many of the pagan gods. It is true that God is often thought of as a father and, in particular, as the father of Jesus. But things really are pretty complex. I should say also that I am not sure that holism is a particularly feminist trait. The Gaia hypothesis, which sees the world as an organic unity, has an early incarnation with the romantics at the beginning of the nineteenth century. The so-called *Naturphilosophen*, or Nature Philosophers, were big into holism—remember, this was the term invented by the South African Jan Smuts in the 1920s. Included in the *Naturphilosophen* were the poet Goethe and the philosopher George Friedrich Wilhelm Hegel. I have never thought of them as much into feminism. Of course, there was Goethe's obsession when he was seventy-four with Ulrike von Levet, who was only nineteen. But I am not sure that that counts!

Fentiman: Right! Right! Can we get back on track, please? I don't know what it is about philosophers that they take the driest topics and find hidden, unflattering depths in them. I think that Lynn Margulis had just invented eukaryotes.

Davies: Hardly invented, but she did tell us how they were first formed. Thanks to some simple cells, prokaryotes, capturing other simple cells, other prokaryotes, we got the formation of complex cells, with interior working parts. Agreed, we are not working with a massive amount of fossil data here, but there are some very strong pieces of confirmatory evidence. Evidence from the rocks, particularly. Free oxygen is thrown into the air by photosynthesis, which takes the energy from the sun and breaks down carbon dioxide. This is one big reason why the destruction of the Brazilian rain forests is a worry. We humans and other animals need oxygen and without plants producing it, there is only trouble. The most primitive producers of oxygen are the cyanobacteria, blue-green algae. These really are halfway organisms between prokaryotes, which get energy by fermentation, and eukaryotes, which get energy by respiration, and, hence, have the need for oxygen. Cyanobacteria have metabolisms on the path between fermentation and respiration, and they appear significantly on the scene around 2.7 billion years ago. There is fossil evidence and even more importantly, first appearing in the rocks at that time, there are organic compounds—compounds akin to cholesterol in us—that only such sophisticated organisms could produce.

There is a big fossil gap now down to 1.2 billion years ago, when we get the first unambiguous evidence of eukaryotes in the fossil record—although it is clear that they must have come after the cyanobacteria. There are major genetic similarities between one kind of eukaryote cell part—chloroplasts, occurring in plants and used for photosynthesis—and cyanobacteria. But, fossil evidence or not, all is not blank. Eukaryotes need oxygen. Do we see a rise in oxygen levels around the time of the appearance of the cyanobacteria, something that continues to a point where oxygen users could thrive? We do indeed! Andy Knoll, a paleontologist at Harvard, points out that iron can rust only where there is oxygen. So, before we get organisms producing oxygen, there should have been no rusting. In a recent book he writes:

> Some of the most compelling evidence for oxygen scarcity on the early Earth comes from gravel and sand deposited by ancient rivers as they meandered across Archean and earliest Proterozoic coastal plains. Pyrite [FeS_2—fool's gold] is common in organic-rich sediments, forming below the surface where H_2S produced by sulfate-reducing bacteria reacts with iron dissolved in oxygen-depleted groundwaters.

The same is true of two other oxygen-sensitive minerals: siderite (iron carbonate, or $FeCO_3$) and uraninite (uranium dioxide, or UO_2). Neither of these minerals is found today among the eroded grains that make up sediments on coastal floodplains, but both occur with pyrite grains in river deposits older than about 2.2 billion years.

Going the other way, after about 2.2 billion years ago, bright red rust containing deposits is common. The Grand Canyon is a great example. "These rocks—called red beds, in the button-down parlance of geologists—derive their color from tiny flecks of iron oxide that coat sand grains. The iron oxides form within surface sands, but only when the groundwaters that wash them contain oxygen. Red beds are common only in sedimentary successions deposited after about 2.2 billion years ago." This doesn't mean that we have oxygen levels like we have today, but they go from about 1 percent to 15 percent, no small jump.

We can keep going with this story, but I hope you are already starting to notice one thing. There are huge numbers of gaps—nobody has a real clue about the start of sexuality, for instance, although it is thought that it truly developed only in the early eukaryotes—but what we do know all fits together nicely. It is a bit like doing a jigsaw puzzle with only a few of the pieces yet in place. But, just like those pieces in place fit with the overall picture—sky is where it should be, and the water mill is where it should be—so, also, the fossil and other pieces fit where they should fit. For instance, fermentation and respiration are not two completely different processes. Respiration is fermentation with more steps added on—just what we would expect if the overall story is true. Cyanobacteria appear long before real eukaryotes. Just what we would expect if the overall story is true. Oxygen appears when it is expected. Just as if the overall story is true.

Moving down quickly now toward the Cambrian, we find increasingly sophisticated organisms in the record. Again, it has to be admitted that there is nothing like the forms we find in the Cambrian, but the fossil remains point to living creatures that are much more complex than those earlier. They are starting to show features like hard skeletons—features that we associate with Cambrian organisms. A major move was the development of bilateral organisms—organisms like humans and horses and butterflies and trilobites that have mirror-image sides. This is obviously a crucial forward step. You could not get a Derby winner with a horse with three legs on one side and only one on the other. Here, evo-devo comes into play. As we have seen, we now know that even fruit flies and humans share very similar sequences of genes. Remember, organisms are built on the Lego principle, where certain pieces are used over and over again. However, the sequences are rarely, if ever, absolutely identical. There are differences between the molecules strung along the DNA macromolecule of humans and fruitflies.

A lot of theory and experimentation has been applied to this sort of issue. It is thought that many of these single molecules are not that central to the functioning of the whole organism and, hence, they change or mutate in a kind of random fashion. Even biologists who are not that keen on genetic drift when it comes to physical features allow that down at this level, drift might be a significant factor. Why not? There is no natural selection working to sift through the molecules and choose some rather than others. This has led to a really important insight. If you have some idea of the rate at which mutation takes place, then you can calibrate a kind of "molecular clock," through which you can measure when two different organisms split apart—the date of their common ancestor.

Of course, the whole thing can get pretty hairy once you get into the technical details, but by comparing different molecules with different rates of change, you can come up with some pretty good measurements. As it happens, in the case of bilateralism, you get just what you would expect—the genes for this fell into place at some point before the Cambrian explosion. The estimated range of possibilities is still pretty large. Not that large, though. No one thinks bilateralism was possible two billion years ago or one million years ago, both of which would be simply devastating findings.

It is true that this still does not in itself explain the Cambrian explosion, but it does show that the parts were in place ready for the explosion, or in less forward-looking language, able to take advantage of whatever it was that triggered the explosion. Some think that the explosion was one simple move forward—a whole book was written recently suggesting that the big breakthrough was the development of sight, which enabled organisms to exploit the surroundings much better. But apart from evidence of pre-Cambrian sight, most people think it was a more complex set of factors. Probably a dramatic rise in oxygen levels was needed, and geology helps here. There is some evidence that the movements of the continents caused massive ice caps and this, in turn, caused the sun's light to be reflected off the earth, causing even more cold. Prokaryotes could no longer ferment, but some photosynthesis was possible, meaning that oxygen levels went up and up. As the earth started to warm again, thanks to the heat pouring out of volcanoes, organisms that used oxygen, the more sophisticated organisms, were ready and able to take advantage of things.

Do you need anything more? A lot of evolutionists think not. A leading University of Chicago paleontologist, Jack Sepkoski, who died tragically young, argued that once the parts were in place, the massive growth followed almost inevitably. There was a huge, empty ecological space, as it were, and for a while the struggle for existence was turned off or at least reduced, and organic numbers and forms shot up. There was a kind of S curve, as things started rather slowly then sped right up and, finally, as the world filled up, the rate of evolution and its change eased off to a kind of equilibrium. This was not just speculation. Sepkoski was part of the first generation of scientists, certainly the first generation of biologists, who was comfortable with computers. He did massive surveys of all of the forms of life that had lived here on earth, and he claimed that one saw this kind of growth, "sigmoidal" growth as it is called, time and again. When the flowering plants came along, for instance, they were able to exploit a huge niche and they did so until it was crammed. In a way, argued Sepkoski, the mistake is to think that the Cambrian explosion was a one-off phenomenon that required a special unique explanation.

Fentiman: Let me try to understand you. Your point is not that the Cambrian explosion was nonexistent or even that it was unimportant, but that by picking it out in the way that he does, Hal Wallace implicitly implies that it requires some special kind of explanation.

Davies: Precisely. If we look at the explosion in context, then it seems a lot less mysterious.

Fentiman: Emily, I can see that you are itching to say something. Let's hand things over to our Episcopalian priest.

Matthews: Yes! You see my kinds of worries are now very obvious. I don't go along with Hal, next to me, in wanting to deny the whole scenario, or even in wanting to suppose miracles for the Cambrian explosion. It is just that there is so much supposition in what David Davies has just told us that I feel that there has to be something more to the tale. I don't in any way feel that I, a practicing Christian, am threatened by the story. If anything, I am more convinced than ever before that my God is at work. The whole story is one of trial and painful progress forward, taking literally billions of years. At one level, it is magnificent; at another level, it is so human. That's why I am drawn even more to the God of Process Theology, who works with us trying to shape and influence things, rather than the Greek God of the early theologians, like Augustine.

Augustine was surely right in seeing God seeding the earth rather than creating miraculously wholesale, but he was wrong in seeing God as all-powerful in the sense that God could do anything all at once. And again, I am worried about relying exclusively on natural selection. It is both too painful and too efficient. Why didn't complex life emerge about three bil-lion years ago? After all, life itself appeared as soon as it possibly could. I much prefer a kind of unfurling. That's why I am drawn to nonselective causes like "order for free" and so forth. The whole story is just—what shall I say—too God impregnated to be pure chance like Darwinians say. At the same time, one does not see the opportunistic efficiency that I associate with Darwinism.

Davies: You know, this is my trouble with you Christians. There is Hal, there, who wants to deny the whole thing—at least, who wants to deny any sensi-ble naturalistic explanation—and then there is you, Emily, who says that evolution is the best of all possible proofs of God. How can you react like this? I'm a Darwinian. I don't have these choices.

Rudge: I'm not sure that that is quite true. You don't have these choices as a scientist—although, note how many gaps you leave open for different hy-potheses—but religiously you do. I accept everything that you say, and like Emily—if in a nonreligious sense—I can only wonder at the glory of it all. What a fantastic tale. As Richard Dawkins is always saying, it is quite equal too—much surpassing—anything you get in Genesis. But, although I don't have Emily's faith, I simply cannot see why it is impossible for someone to ac-cept this story and still think that Jesus died for our sins. You can't do so if you are a literalist like Hal, but we have already been over the ground of how much a traditional Christian is committed to the literal truth of the Bible.

Of course, where I differ from Emily, and am right with you, Dave, is over the selection issue. Like you, I just don't see the need to invoke any other significant mechanism. Like you, I don't think that Emily really un-derstands the real nature of science. It is not a matter of having all of the solutions at once. It is much more a matter of filling in the pieces. Didn't you compare it to doing a jigsaw puzzle? That is what you scientists are do-ing, and what you said about the origin-of-life problem is very pertinent here. It is only recently that we have developed the really powerful tools to look into these questions—the evo-devo people and their molecules, Jack Sepkoski and his computers—and why should we not stop now or look for other solutions? The study of life is going forward at an incredible rate. Now is not the time to cry "failure."

Fentiman: We need to keep moving on. Let's suppose that we have got rid of the dinosaurs and we are now in the age of mammals, as it is called appro-priately. I understand—and, again, I accept that Pastor Hal will dispute this—that primates, monkeys and those sorts of things, first arrive on the scene about fifty million years ago. What's the picture here? I'll turn first to our resident evolutionist to get us going. Professor Davies?

Davies: The problem is that if we humans evolved from monkeys—not monkeys of a kind that we have around today, obviously—then these ancestors of ours probably lived in the tropics in jungles and the like. That's where we find monkeys today, so why not then? Unfortunately, jungles are just about the worst place for fossils and, expectedly, we don't have much evidence. But, now we are starting to dig up pieces of bone from the last twenty million years or so, mainly in northern parts of Africa. However, you have got to remember what I was saying earlier about inferring past histories, phylogenies. Fossils are important, but they are not the only evidence. We see this point in spades here. Thirty or forty years ago, if you had asked people, the kind of people who are interested in human evolution— they are known as "paleoanthropologists"—about our origins, they would probably have said that the human line broke off about ten or more million years ago. Everyone thought that we are completely distinct from the animals that are surely our closest relatives, the chimpanzees and the gorillas. After all, we are different. Big brains, walk on two legs, smooth, hairless skin, and so forth. Then, the molecular biologists got into the act and said that we are so close to the great apes that we cannot have split from them much more than five million years ago, and—oh, by the way!—we humans are closer to chimps than chimps are to gorillas! Our line split from the gorillas and only later from the chimps. Of course, the bone people were outraged and denied that this could possibly be true. But, the molecular evidence kept coming in, and now no one denies it.

In other words, as I have been saying all along, there is more to inferring the past than digging up bones. This is not to say that the fossil evidence is unimportant or that people no longer search for it. You often hear talk about missing links and how evolution will always be a bit iffy without them. In fact, the missing links in human descent—"human phylogeny," as it is called technically—are not really so very missing. At the time of Darwin, people had started to discover Neanderthals, those primitive, like-humans who used to roam Europe. General opinion—back then and it's still true today—was that the Neanderthals were not a different species from us. A lot of people, in fact, thought that they still live in Ireland!

The first real prehuman fossil was found by a Dutch doctor, Eugene Dubois, searching in the East Indies. "Java man," as it was called, is a genuine specimen of an earlier member of the genus *Homo*, to which we belong.

Fentiman: Would you mind spelling out for our viewers what all of this means? What is a genus and so forth?

Davies: No problem! In the eighteenth century, the Swedish biologist Linnaeus devised this hierarchical system, so that every organism belongs to a group, known as a "taxon," at each level. The levels are known as categories. Linnaeus had seven levels, although today, we often have more. The names are always in Latin, although they are often made up and a bit silly. If we found our host in the fossil record, we might call it *Homo redversfentimanus* or *Homo eternalquestionsus*.

Fentiman: I'm not sure that this is so silly, but go on.

Davies: So, how does the system work? Take us humans, for instance. We are members of the taxon Animalia at the highest category level, the kingdom. Sunflowers are members of the taxon Plantae. Going down the levels, we humans are next in the group of animals with backbones and a few others with a kind of primitive backbone. This is the taxon Chordata, and it occurs at the level, the category level, of phylum. Then, there is the category level known (perhaps a bit misleadingly) as class. We belong to the Mammalia. Getting more familiar, we have the category of order, where we are Primates, and then the category level of family, where we are Hominidae. Finally, the category of genus, we are *Homo*, and the genus of species, we are *Homo sapiens*. Somewhat inconsistently, the category name and the species name are italicized, and the species name is two words, beginning with the generic name. If you want to be really fussy, you always capitalize the first letter of the genus name, but not the second word that gives you the species.

The great thing about the Linnaean system is that it enables you to find something at once, to tell its relationships to other organisms. In the *Origin of Species*, Darwin showed that the system reflects history. So, you can look up features and work out evolutionary relationships. If you know, for instance, that human beings are Mammalia, then at once you know that, like all Mammalia, or mammals, we are warm blooded, we bear live young, we suckle our young, we have bodily hair—even if not as much as chimps!—and so forth. We are also more closely related to other Mammalia, like cows and sheep, than we are to Reptilia, like lizards and dinos. We now classify Dubois's Java man as *Homo erectus*, reflecting the fact that it is very like humans, but not quite. The brain is smaller. But Java man is a closer relative to us, *Homo sapiens*, than *Drosophila melanogaster*, a fruit fly, or *Gorilla gorilla*, the gorilla, for that matter.

Continuing the story of human discoveries, in the early twentieth century—actually, 1924—a South African anatomy professor, Raymond Dart, found the skullcap of a being that is surely ancestral to us, but not in the same genus. This was a member of what is known as the genus *Australopithecus*, or "australopithecines" in everyday language. As a matter of fact, for various reasons, people did not get too excited about this at the time. Partly because they were looking for human ancestors in Asia and not Africa.

Wallace: Oh, you're being a little disingenuous here, aren't you, Dr. Davies? What about Piltdown man? What about the discovery—or should we say "discovery" in quotes—of the ape-human in England around the time of World War I? What about everyone thinking that humans evolved in the British Isles? And what about the fact that this, like so many great evolutionary studies, proved to be completely fraudulent? That the skull of Piltdown man was really part human and part stained orangutan? No wonder it looked like a cross between a human and an ape. It was!

Davies: True, true, I am afraid. Although it was also scientists at the British Museum in the 1950s who discovered that it was fraudulent, and detection methods are far more sophisticated today. It couldn't happen again.

Wallace: Couldn't happen again? Well, what about the hobbit?

Fentiman: The hobbit? Surely that is a story, *The Lord of the Rings*? No one thinks that Frodo and Sam and all of the others are really real.

Davies: We are not talking about quite the same thing. Let me finish and I will explain. I can do this quite quickly. The main point is that in the twentieth century, especially the second half, huge numbers of fossils have been found, enabling us to trace with some considerable accuracy the course of human evolution. Everyone knows about the Leakey family who have been so active and successful in Kenya in West Africa, but Ethiopia has, if anything, been even more fruitful. Probably the most important fossil find of all was Lucy, or *Australopithecus afarensis*, found by the American Don Johanson. We now think she is over three million years old and the incredible thing is that she was about three feet six and walked. Probably not as well as we do. She was still good at tree climbing, but she really was up on her back two feet.

And more than that. She had a tiny brain about the size of a chimpanzee—that is, about 400 cc. I want to stress that she did not have a chimpanzee brain. She had a brain the size of the chimp's. After her—and the evidence is that she really was female—brains grew bigger and bigger until they were about the size of ours, 1,200 cc. Actually, the Neanderthals probably had brains slightly bigger than ours. But, that is not necessarily really significant. Humans are sexually dimorphic—there are differences in our sizes with males being a bit bigger, including brains. It doesn't follow that males are brighter than females.

A couple more things: well, three things if you include the hobbit. First, why did it all happen? There seems to be no doubt that the big move was leaving the jungle and going out onto the plains. Why we left the jungles is a good question, other than that they were drying up because of climate changes. Why us? Perhaps we just lost the fight with the chimps and gorillas. We came second and had to leave. On the plains, walking properly rather than knuckle-walking like the apes has advantages. It is less tiring if you are covering distances. You are always up looking around for danger and opportunities. There is less sun overall on the body, so that is an advantage in Africa. The most important thing is that the increase in brain size required a steady diet of high-quality protein. Cows have small brains. Carnivores, meat eaters, have bigger brains. Probably at first we were like jackal primates. Grabbing the leftovers from the big predators like lions. But, as we got bigger brains and cleverer, no doubt we learned how to work together like wolves and look after ourselves.

Second, since someone is bound to ask. What about the Neanderthals? Especially if they had bigger brains than we do. It is thought that today's humans came out of Africa about 140,000 years ago, hitting Europe and going west, all the way to Australasia.

Wallace: Ah, Mitochondrial Eve! Clear proof of Genesis.

Fentiman: Eh?

Davies: The Reverend Hal is referring to the fact that, by looking at certain cell parts and the ribonucleic acid that they contain—the things we were talking about last week—we can work out relationships, using the molecular clock, more exactly than we can by looking at the main pieces of DNA. The cell parts called "mitochondria"—these are the power plants of the cell—have their own DNA, and are almost certainly the descendants of separate prokaryote cells that, as Lynn Margulis argued, were absorbed by other cells. Interestingly, for some reason, the mitochondria in a cell are almost always inherited from the mother and not the father. Using this fact, about twenty years ago, three scientists worked out that all humans today have the same ancestor, a female who lived about 150,000 or more years ago. They called her Mitochondrial Eve and, expectedly, the newspapers picked this up and some touted it as the proof of Genesis. We all come from the same universal mother.

Actually, what was proved was not at all that we had one and only one shared ancestor. There might have been a whole group of Eves and Adams who were ancestors to us all. Suppose there were four universal ancestors, call them Adam, Bertha, Christopher, and Deirdre. Adam and Bertha have kids, including the boy Frank. Christopher and Deirdre have kids, including the girl Eve. Already the mitochondria of Adam and Christopher are gone. Now, David and Eve have a kid. From now on, all of the mitochondria are those of Eve, although she is descended from the four people. There might be lots of others in the population, and they, too, are contributing, but by the vagaries of reproduction, their mitochondria also get eliminated. It is like surnames, and the way in which they get eliminated so that, as in Wales, everyone is called Evens or Jones or Davies. So, there is no reason to think that there was one unique woman who alone gave birth to the human race. In any case, she would have been well over a hundred thousand years ago, far too old for the Reverend Hal.

Rudge: To put in another, more general point, I think it is always very dangerous to pick on some new scientific idea and to say that this proves the Bible. One of the popes did it in the middle of the twentieth century, claiming that the big bang proved Creation. Apart from anything else, science is always being revised and you are liable to be left with egg all over your face. But as you know, I would say that science and religion should not be brought together in this sort of way in the first place. To talk of God as Creator is to say nothing about the big bang. It is rather to say that God is the sustainer of existence. Given that God is eternal, the universe could well be infinitely old as we measure time.

Fentiman: Back to Neandertals!

Davies: The Neanderthals, who were in Europe already, probably got wiped out, although whether through violence or simply through not getting the best places to live is hard to say. Probably there was some interbreeding with modern humans, although probably not much. It has been suggested that perhaps the Neanderthals could not speak properly like we can, although most people doubt that. They had some culture, like burial patterns, but, overall, one suspects that they were simply not as sophisticated as modern humans. About fifty thousand years ago, we were in charge, although it was probably another forty thousand years before the idea of agriculture was developed and the path was cleared for life as we know it today.

Finally, hobbits. Well, they are not really hobbits. About five years ago, on the island of Flores, part of Indonesia, some Australian researchers found remains of these little people, about three feet tall, who seemed humanlike, but not humans as we know them. They weren't simply pygmies of the sort that we find in Africa. There has been—there still is—a lot of debate about them. They are known as *Homo floresiensis*, but that is a bit of a guess at the moment. At first thought, the feeling was that these were shrunken versions of recent prehumans, like *Homo erectus*, the species from which we are believed to be descended. That is not such a silly idea. On islands, organisms often evolve in the direction of smallness. There is not as much food around, nor predators, that make size a major adaptation. On Flores, there are fossil remains of dwarf-sized elephants.

However, judging by bones and so forth, some people now think that *Homo floresiensis* may be more distantly related to us, perhaps have australopithecines as their more immediate ancestors. I should say that there is a minority that thinks that hobbits, as everyone calls them, are simply humans with genetic problems, namely small brains. The hobbits truly were human morons. This denial is only a minority position, though.

Fentiman: Well, I think we have enough on the table to get going into discussion of the relevance of all of this to religion.

Davies: Yes, if I can finish. It seems to me that the story I have just told blows religion out the window. If you accept the evolutionary story of human origins, then Genesis is simply false, and there is just no way that one can say that humans are made in God's image. We are animals just like all of the others. It's religion or science, but not both, and science wins. A conflict, and now it's over.

Matthews: I couldn't agree with you more! I couldn't disagree with you more! As a Christian, I have absolutely no problems whatsoever with what you have just said, Professor Davies. I am glad to hear all of the details. If God wanted to create in an evolutionary fashion, then that is just fine by me. As I have said before, for myself, I think there are good theological reasons for a lawbound world, quite apart from the fact that I am not sure what a nonlawbound world would be like. Also, perhaps more importantly, I am with the great nineteenth-century Danish thinker Søren Kierkegaard in believing that God's presence speaks to you all the time in the world, but you cannot prove anything. In an important way, the whole point of the world we have is that we have to have faith. There are no QED answers, like you get in mathematics.

But as I have also said, unlike Professor Davies, who believes that science and religion are in conflict, I am an integrationist. I think that science and religion speak as though with one voice. Of course, there are some things that science tells religion. About the hobbit, for instance. Don't go searching the Bible for evidence of them. We mentioned Genesis, chapter 6, which speaks of there being "giants in those days." Hal thinks this a scientifically verifiable fact. I could not disagree more. That is not a scientific statement or an invitation to go to Africa to look for bones. As a Christian, I could not care less about those giants, real or imaginary. Of course, there are some things that religion tells science. About immortal souls, for instance. There is something divinelike about us all. Not God, but the image of God. Quakers speak of "that of God in every person," or the "inner light." There is something not material that gives hope of eternal life. This is not a scientific belief, and people like Richard Dawkins who think that unless something is scientifically provable, it is false are just naive. All of this said, what I believe as a Christian, and what Professor Davies has just told me as a scientist, go together perfectly.

Fentiman: Thanks for that.

Matthews: Hang on a moment, there is a bit more that I want to say. I'm an integrationist. I don't think you should search science for religious evidence, nor do I think you should search religion for scientific facts, but I do think that truth cannot be opposed to truth. I expect in some sense to see the one picture reflected in, harmonizing with, the other picture.

Rudge: I am just not sure that that is a consistent position to take. I think you are playing fast and loose with harmonizing and so forth. If you like the facts, you can use them. If you don't like them, you can ignore them. Better to be like me, and keep the two pictures strictly apart.

Matthews: And I in return don't think that you are being consistent. If you are prepared to allow that religion and science can both be true—and that seems to be your position, even though you yourself are a nonbeliever—then I just don't see how you can claim that they never interact. But rather than just squabbling about the theory, let me tell you how I see the human question in the light of religion, given the science that David Davies has just given us.

For a start and most importantly, we humans are not just any other animal. We are the end point, the apotheosis, of the evolutionary scheme. Here we are now, with consciousness and awareness, with intelligence and a moral sense, like no other animals in creation. Julian Huxley, Thomas Henry Huxley's grandson, who was an out-and-out atheist, used to say that we humans are so distinctive that we constitute another kingdom, along with animals and plants. Incidentally, he was an enthusiast for the French Jesuit paleontologist Pierre Teilhard de Chardin, who argued in his book *The Phenomenon of Man* that all progresses up to the Omega Point, to Jesus Christ. A lot of people have criticized Teilhard because he was not a Darwinian, but—apart from the fact that, as you know, I am not that keen a Darwinian myself— I believe we can bring Teilhard up-to-date in the way that Professor Davies has just sketched. Humans are special, we are the ones with the supreme attributes, and we come at the end of the process. For me, that is the real meaning of the Creation story of Genesis. Not that we were made miraculously on the sixth day, but that we humans are the climax of God's creation.

Davies: From humans to Teilhard de Chardin! From the sublime to the ridiculous! Sir Peter Medawar, the Nobel Prize winner in physiology, put his finger right on it and similar lunatic systems. He said that Teilhard's thinking is "nonsense, tricked out with a variety of metaphysical conceits, and its author can be excused of dishonesty only on the grounds that before deceiving others, he has taken great pains to deceive himself."

Rudge: Far be it for me to speak up on behalf of a Jesuit, but I can't help feeling that Medawar was being a bit mean here. I agree that Teilhard's position as he presented it just won't do. He did indeed reject, or at least ignore, Darwinism, and there was more than a whiff of Bergson's vitalism about his system. On the other hand, it seems to me that what he was trying to do was what a Christian who is a scientist should be trying to do. He was trying to take seriously his science and his religion. The trouble was that he claimed that what he was doing was pure science, and it obviously was not. Not only was Teilhard in trouble with the scientists, from his perspective more importantly, he was in trouble with his church. The Catholic authorities thought he was being heretical. To get around their objections, he claimed that his theory had nothing to do with religion and was pure science. Unfortunately, his efforts were in vain because the authorities banned his work, and his ideas were only published after his death.

In a way, though, more important than Teilhard's troubles are the underlying assumptions of his thinking. We should turn to the issue of progress. Emily staked a strong claim about humans and progress. She argued, like her inspiration Teilhard de Chardin, that the evolutionary process is progress, starting with the blob and going up to humans. "Monad to man," they used to call it in the old days. Putting on my historical hat, let me say simply that right from the beginnings of evolutionary thinking in the eighteenth century, people identified organic change with progress. Just as in the human world, you get more technology and more education and better health care, so in the organic world you get a steady march up from the very primitive organisms to the very complex, and finally—Ta da! Ta da!—you get humans. A lot of people still believe this today. Today's most eminent evolutionist is Edward O. Wilson. We mentioned him before. He is a Harvard professor and the world's leading expert on ants. He writes on a lot of other things as well, including conservation and human biology, or rather human social biology. Wilson is quite committed to the belief that we see an upward drive in evolution.

The trouble is that, in respects, Darwin—who, incidentally, himself believed in biological progress—knocked the bottom out of all of this. If the evolutionary mechanism is natural selection, then what does this mean? It means the fittest survive. But who are the fittest? In one situation, it might be one organism, and in another, it might be another organism. Remember the hobbit. Is it better for humans to be big or small? It all depends! For us in Europe and America, biggish is good. On the island of Flores, it seems that small was good. This doesn't mean that humans aren't better than other organisms—I would rather be a human than anything else—but it does mean that evolution doesn't prove this. Add to this the randomness of the building blocks of evolution—the mutations of the genes are random, not in the sense of being uncaused, but in the sense of not appearing to order—and it seems that evolutionary progress is a hangover from the eighteenth century.

Stephen Jay Gould was one who was strongly against the idea of biological progress. He said—I thought this would come up, so I'm prepared—that progress is "a noxious, culturally embedded, untestable, nonoperational, intractable idea that must be replaced if we wish to understand the patterns of history." Remember that comet that wiped out the dinos. Gould again: "Since dinosaurs were not moving toward markedly larger brains, and since such a prospect may lie outside the capabilities of reptilian design . . . we must assume that consciousness would not have evolved on our planet if a cosmic catastrophe had not claimed the dinosaurs as victims. In an entirely literal sense, we owe our existence, as large and reasoning mammals, to our lucky stars." The simple fact is: there ain't no progress. Organisms live and then they die. End of the matter. Somewhat meanly, Gould even went as far as to accuse poor Teilhard of being the perpetrator of the Piltdown hoax—he wasn't, I should say—so that he could tar the biggest progress booster of the twentieth century.

Having said this, a number of biologists, good Darwinian biologists, would disagree. They think that you can still get progress from natural selection. Richard Dawkins, for instance, seizes on the notion of biological arms races. Remember how this suggests that, in evolution, the prey gets faster to escape the predator, and the predator gets faster to catch the prey. Dawkins thinks that in military arms races in the twentieth century, we have seen a move from heavier armament and stronger guns and bombs to more sophisticated electronic equipment, and in the living world, he thinks that this means that there is bound eventually to be a move to brains. Big brains like humans. The Cambridge University paleontologist Simon Conway Morris has a different idea. He thinks that there are certain ecological niches, as it were, waiting to be occupied by organisms. Natural selection is always looking for these and pushing life into them. So, for instance, there was a niche for a tigerlike being with saber teeth, and at least twice this niche was occupied—once by marsupials and once by placental mammals. Conway Morris thinks that there is a kind of succession of niches, ending with one for intelligent beings, and so progress was pretty much bound to happen.

Myself, I think there are problems with both of these approaches. Of course, humans evolved, so there must be some cause. But, it seems to me that Steve Gould may have been right when he said that if life can get more complex over time, then it simply has to get more complex over time, purely by chance. It is like a drunkard walking down a sidewalk, with a wall on one side and a gutter on the other. Eventually the drunkard will end in the gutter, not by design but by chance. He can't go through the wall!

The trouble with arms races is that there simply does not seem any necessity for brains, especially big brains, to occur. Being big brained from a biological point of view is not always a good thing. Apart from anything else, big brains demand lots of protein, and, as we have seen, that means meat. There were not many vegetarians in the Pleistocene. But getting meat is a big job and not always worth the effort. If there is nothing but grass, better to be a cow. The trouble with Simon Conway Morris's idea is, first, that it is by no means obvious that niches just exist independently of organisms, waiting to be conquered. One major niche for insects is the canopy of trees in the Brazilian jungles, but that niche only exists because the trees exist. Second, and a much bigger problem, is that Conway Morris seems to assume that there is an ordering of niches, with the intelligence niche top. Even if he proves that we had to evolve, he doesn't prove that we are thereby better.

This is why I want to go back to my independence position. I agree with Emily—and with Pastor Hal, for that matter—that if you are a Christian, you have to believe that humans are special. Not necessarily unique. Christians have divided over whether there could be intelligent life elsewhere in the universe. But somewhere, somehow, something humanlike had to appear. Perhaps with green skin and six fingers, but you have to have something with intelligence and a moral sense. I am not sure whether you have to have sex, although it might be the case that you couldn't get sophisticated organisms like humans without sex. Notice, I'm not saying we're not sophisticated. The question is whether we are better. The trouble is that I am not convinced that you can truly get progress out of Darwinian evolution. I am speaking scientifically now, although, speaking philosophically, it seems to me to try to get organisms, valuable in some absolute sense, from a blind process is trying to get values out of science.

Personally, if Gould's randomness argument does not work—and it might work if you allow God the freedom to have life all over the universe, so one case will surely strike pay dirt, and if not in this universe, then in some other—I would rather simply separate science and religion. Emily wants to use science to bolster her religion. I want to say that humans evolved and that is enough for religion. We know already that Saint Augustine argued that God is outside time, so for God, the thought of creation, the act of creation, and the product of creation are as one. Creating the world, he knew that humans would appear. Perhaps that makes it all seem a bit deterministic, but then Augustine is a bit deterministic. An omnipotent god knows how things are going to turn out.

Although, to pick up on the point that I just made about God's freedom, do remember, if God does stand outside time—"one day is with the Lord as a thousand years, and a thousand years as one day" (2 Peter 3:8)—then he can keep creating universes over and over again, without any sense of this taking trillions and trillions of years. What might be trillions in our eyes is the same instant for God. We know that the Darwinian process can lead to humans—it has done so in our case!—God knows this, and so the randomness of the evolutionary process is no big deal for him.

Davies: Yes, but you only get to this conclusion by making all sorts of silly assumptions, particularly about God being outside time. You can't be outside time. That is the truth of the matter. We are born and live and die. Same for planets and stars. If wishes came true, then beggars could ride. Wishes don't come true, and nothing exists outside time. Trillions of years are trillions of years, not an instant.

Matthews: Perhaps I can jump in here. It seems to me that the fundamental question of philosophy—not of faith—is the question posed at the beginning of his *Introduction to Metaphysics*, by the great twentieth-century German thinker Martin Heidegger: "Why is there something rather than nothing?" A lot of Anglo-Saxon philosophers responded that you cannot answer that question, so it must be meaningless. I am not sure that that is much of a response. The skeptic, I guess, says that you simply cannot answer it and leaves things at that. The theist, the Christian, says that there is something because God made it. But, then there is the question of what made God. People like Richard Dawkins seem to think that this is the end of matters, and that you are simply caught in an infinite regress and that is a refutation of the whole God hypothesis. The Christian says that nothing made God because God is not something that needs making. Unlike us, who are contingent, God exists necessarily. That is why God exists outside time. We talked about this back in the first program, but it's worth reemphasizing. Two plus two never became true and it will never cease to be true. Likewise with God.

Oh, and incidentally, I don't agree that trillions can't be an instant. We've all had the experience of falling asleep and then the next moment it's morning, even though we know that several hours have passed. God's not asleep. Anything but. However, time is as unreal for him—or let us rather say, non-binding on him—as it is for us or on us when asleep.

Rudge: I think you should point out, Emily, that not everyone finds the idea of a necessary being entirely coherent. We have already made mention of the late-Enlightenment philosopher Immanuel Kant, who denied that existence was a property like red or round. He would have had trouble here. Also, you should mention that even though this may be traditional theology, you yourself reject it because, as a Process Philosopher, you think that God is in time, working along with humans.

Matthews: Agreed, although to be fair, don't forget that all Christians believe that God, however eternal he may be, was prepared to come into the temporal frame through his son, Jesus Christ. You're right that I myself want to make God much more part of the picture, but then don't forget either that I think that evolution is probably more guided, more directional, than someone like Professor Davies believes. So, I'm not stuck with the paradox that Christians insist that humans or humanlike beings must appear and yet the evolutionary process is essentially random. I think humanlike beings must appear, but I don't think that the process is as random as is suggested by pure Darwinism.

Fentiman: And on that note, I am afraid, we must bring this evening's discussion to an end. On behalf of our panelists, this is your host, Redvers Fentiman, saying good night. Join us next week for the final program in this series of *Eternal Questions*.

CHAPTER FIVE

~

Program Five: Humans

Fentiman: Good evening. We have arrived at the fifth and final program in this series of *Eternal Questions*. Our topic is the relationship between science and religion. Tonight, we are going to close with a look at the most important issue of them all, namely, us! We are going to ask about the status of humankind, as viewed through the lens of the science-religion debate. My panel includes the usual suspects. From left to right, we have Professor David Davies of the Massachusetts State Institute of Science. Then, we have Professor Martin Rudge. He teaches the history and philosophy of science at Robert Boyle College in Minnesota. On my right is the Reverend Emily Matthews. She is an Episcopalian priest and teaches counseling at Wycliffe College. And next to her is Reverend Harold Wallace, head pastor of Rollingbrooke Stones, a large Baptist congregation in Atlanta. Welcome, everyone, and welcome to our studio audience and to you at home.

We talked last time about human evolution. Tonight, I'd like us to start by focusing on humans as they are today, and what we know or think about them from a scientific perspective. How this impinges on religion, if indeed it does. I suggest that we start with biology and then, perhaps as is pertinent, bring in the social sciences. I would like to leave time at the end of the program for everybody to sum up their personal position on the science-religion relationship. Instead of asking our scientist, David Davies, to kick off our evening's discussion, I am going to turn to our philosopher, Martin Rudge, because I know he has written on the subject. Over to you, Martin.

Rudge: Thanks. I should say that I have written on it because, religion quite apart, it has been a topic of considerable controversy, especially in my own field of philosophy. Let me try to lay out some details, without at first revealing my own hand, although that will become apparent fairly quickly. The easiest place to begin is with *Sociobiology: The New Synthesis*, a great big book published in 1975 by a man whose name has come up before, the Harvard ant expert, Edward O. Wilson. Today, thanks to a great passion that he has for saving the Brazilian rain forests, Wilson is the darling of the environmentalist movement, but thirty years ago, he was something else. At least, he was considered something else. In the last program, mention was made of the fact that Wilson has looked at various things, not just ants, but also human social biology, and it is this latter that was the center of a huge controversy, or row.

In a sense, it was all a bit silly, or accidental. By the mid-1970s, evolutionists had really moved in a big way into explaining social behavior—the ants and the birds and the fish and the mammals, and especially the apes. What Wilson did was try to gather everything together in one big picture, and call it "sociobiology." I don't think he invented the name, but he used it in a flamboyant and rather provocative fashion, starting with making it the title of his book, an overall survey of the field. The trouble was that the final chapter of the book turned to human beings, and Wilson argued that we, too, are part of the evolutionary picture. The things we do, the things we think, the things we want. He elaborated on this two or three years later in another book, *On Human Nature*, which went on to win a Pulitzer Prize.

No harm in any of this, you might think. Well, in the opinion of many, you might think wrong. Wilson argued, for instance, that male-female differences in humans—males big and strong and aggressive, females soft and tender and coy—is part of biology. Males have been designed by natural selection to be what they are. Females have likewise been designed by natural selection to be what they are. He said that things like ethics, our sense of right and wrong, are just part of our biology. Notoriously, a few years later, in a piece he wrote with the philosopher Michael Ruse, he referred to morality as "an illusion of the genes to make us good social animals." He argued that aggression was part of our innate nature, and that this applied especially to the fear that we have of strangers. Being nasty to Jews is part of the in-group/out-group legacy that we have from the Pleistocene past, when we humans lived in groups and the biggest threats came from other groups of humans. Even religion was part of Wilson's picture. He thought it came about because it promotes group cohesion, but has no real meaning at all. Once it is explained, it is seen to be total nonsense.

Hang on, I have a quote here, because I thought it would come up. It's from *On Human Nature*. "As I have tried to show, sociobiology can account for the very origin of mythology by the principle of natural selection acting on the genetically evolving material structure of the human brain. If this interpretation is correct, the final decisive edge enjoyed by scientific naturalism will come from its capacity to explain traditional religion, its chief competition, as a wholly material phenomenon. Theology is not likely to survive as an independent intellectual discipline."

Wallace: I am not surprised that Wilson was controversial. At least 80 percent of Americans believe in a personal god, and most of the rest believe in a nonpersonal god. Here is a man denying the whole thing and saying that it is part of our genes.

Matthews: Goodness, Hal! Surely you are not going to take that sort of thing too seriously. We all know that when scientists start to make philosophical or theological inferences from their work, it is almost a sure guide that their conclusions do not follow! There seems to be a conviction, on the science side of campus, that philosophy and theology can be done after a couple of drinks in the faculty club on Friday evenings. It's almost as bad as when retired scientists start to write histories of their heroes. Isn't there a joke term about scientists getting old and entering the "philosopause," when they start writing about God, Truth, and the Meaning of Life?

Davies: Thanks a lot!

Rudge: Actually, it wasn't the Christians who got really tense about human sociobiology. A lot of social scientists didn't much like it. I guess that was to be expected. They saw their turf being invaded by the biological barbarians and they felt threatened. More interesting and perhaps surprising, a very vocal subclass of biologists objected strongly to human sociobiology. What made the whole thing rather nasty and personal was that two of the leaders—the paleontologist Stephen Jay Gould and the geneticist Richard Lewontin—were in the same department at Harvard as Wilson.

Matthews: Didn't we talk about these people last week or the one before? Weren't they the people who wrote the paper about spandrels, the triangular bits at the tops of columns in medieval churches? They argued that spandrels really have no function, they are just by-products of the building process, and that this is true of a lot of organic features. They are just by-products and we shouldn't keep looking always for adaptation.

Rudge: That's right. I think their attack on adaptation—at least their attack on ubiquitous adaptation, what they called "pan-adaptationism"—was truly an attack on human sociobiology. As Wilson read it, nigh everything about humans was biologically adaptive. If Gould and Lewontin could destroy the general assumption that biological adaptation is universal, then indirectly, they destroyed human sociobiology. Why they wanted to do this is another matter. Both of them, at the time at least, were Marxists and they disliked any explanation of human nature that was not couched in terms of power or class struggle.

Davies: I heard the old evolutionist Ernst Mayr once say that he thought a lot of the opposition to human sociobiology came from American Jews, and this of course included Lewontin and Gould. They feared any system that supposed that biology determined human nature. Speaking myself as a Jew, I think their fears were overblown—I don't see that it is anti-Semitic to say that people learned to dislike Jews because their biology predisposes them to hate outsiders. On the other hand, I confess I have a certain sympathy for their stand.

Fentiman: Martin, just finish up, will you? You've been talking about thirty years ago. Where do we stand today?

Rudge: In many respects, things have grown a lot more quiet. One thing is that human sociobiologists tend not to call themselves that anymore. They generally go under the label of "evolutionary psychologists," and apparently that is a lot less threatening. I wouldn't say that there is a large number of them, but those that there are work away vigorously on various problems. For instance, there was a fascinating study by a couple of Canadian scholars that showed that stepparents, stepfathers particularly, are much more prone to violence toward the kids than are biological parents. Biologically, this is not surprising. In the animal world, the new male in a family often kills off all of the kids from the previous male. This brings the female back into heat and ensures that she is ready to look after his offspring. In humans, this was such an unexpected finding that until the Canadians made their hypotheses public, social services and police forces had never even thought to distinguish between stepparents and biological parents in cases of family violence.

Of course, a lot of people still don't much care for evolutionary psychology, or whatever it is called. Cultural anthropologists tend to be somewhat unbalanced on the topic. Feminists also go bonkers when thinking about the subject. The very suggestion that male-female roles might have a biological underpinning, rather than being a social construction forced on women by bullying males, is enough to bring on spasms of apoplexy. My own field of philosophy is interesting. Initially, there was a huge amount of hostility to human sociobiology. When it comes to God issues, philosophers tend to be pretty agnostic like me. But, they still don't much care for suggestions that we humans are really at one with the animals. We laugh at Descartes, but in our hearts, we think he was right in putting humans in a special place apart from the animals. A lot of philosophers were influenced by Lewontin and Gould. Today, a good number still think that way. However, particularly in ethics, there is a lot more sympathy for a biological approach.

Matthews: Perhaps I could get in on the discussion here. The really interesting sociological fact, I guess you could call it, is that although feminists and social scientists and philosophers didn't like human sociobiology, we liberal Christians just loved it. We really thought it was the cat's pajamas. This may, of course, have been part of our eagerness to show that we are onside with science, but I think it was something deeper. The whole issue of morality—Why should we be good?—is something that we wrestle with, and it seemed to many of us that a biological approach was like a breath of fresh air. In this respect, I think we beat the philosophers to it.

Rudge: I will go along with that. Some philosophers saw the potential right from the beginning, but it is only in recent years that there has been a tsunami of interest in the topic.

Fentiman: Emily, since you Christians apparently beat the philosophers to it, why don't you tell us a bit about things. What is the biological contribution to the understanding of right and wrong? I would also like to see how this ties into your integrationist position.

Matthews: Why are we good? Pick up on a concept that was mentioned in an earlier program. Richard Dawkins has said that genes are selfish, but he does not imply that humans are necessarily selfish. Genes work to replicate themselves, and the better they work, the more they replicate. That's the sense in which they are selfish. It's a metaphor. Sometimes the bodies carrying genes—organisms, animals—can do a better job of gene replication if they cooperate with others than if they just fight flat out. Suppose you have two animals that find a cache of food—say, jackals finding the remains of a lion's meal. Instead of fighting for all of the food and perhaps losing it all to a third animal who sneaks it away during the fight, better to cooperate and share the cache. Half a carcass that you do get is better than a whole carcass that you don't get. Correct me if I'm wrong, Professor Davies, but I think this is known as "reciprocal altruism." You scratch my back and I'll scratch yours.

Davies: You are absolutely right. We'll make a biologist of you yet, Emily! Of course, it is not the only way you get cooperation and organisms helping others. Parents looking after children is the most obvious example. There is no point in having lots of kids if they die off immediately or before reproducing in turn. So, looking after your offspring is something that can be promoted by selfish genes. More generally, helping relatives reproduce can be a good evolutionary strategy. Relatives have genes of the same kind, so, in a way, it doesn't matter if copies of my genes are passed on by me or by relatives. Normally, of course, you pass on more of your own genes if you do the reproducing yourself, except for identical twins, who have exactly the same sets of genes. You're only half-related to your brother, for instance. But, if you can get him three kids in exchange for one of yours, that is a good deal.

Fentiman: Sorry, Emily, but I still don't see why someone like you, who is a Christian and believes in God, and at the same time is an integrationist, would welcome human sociobiology. As far as I can make out, if the biologists are right, then morality is nothing more than an adaptation to make us social. It has no deeper meaning.

Rudge: If I can jump in here, that is precisely the point. David Hume said that morality is nothing more than a matter of sentiments, feelings. Hume also said that there is a difference between claims about matters of fact and claims about matters of morality. You cannot go logically from the one to the other. I may dislike rape, but it doesn't follow from that that rape is wrong. I dislike Bud Lite, but—hard as it may be to grasp—it does not follow that a Bud Lite drinker is immoral. We usually say that Hume's point is that you cannot go from "is" statements to "ought" statements.

That's one of the reasons why I don't much like biological progress. Too often, people—Ed Wilson is a paradigm example—go from talking about the path of evolution to saying that the winners are better. All that leads straight to moral prescriptions. Humans are the top dogs. Hence, we should do everything in our power to look to the interests of humans. Don't introduce genetically modified foods because they are not natural—they did not evolve. And so forth. I am not sure that Wilson was very consistent on this, but that is why the philosopher Ruse wanted to say that ethics was an illusion. Obviously, he wasn't saying that rape is okay. What he was saying was that there is no ultimate justification for condemning rape. He was saying that our feelings that rape is wrong are adaptive sentiments, put in place by natural selection to make us good social beings.

Matthews: And I want to go right along with that. It's just that as a Christian, I don't think Ruse and Wilson go far enough. In the context of science, you don't get a foundation for ethics. You just get feelings. But, I as a Christian know that there is something more. I know that God wants me to behave in certain ways. He doesn't want me—or rather you—to rape, because it is wrong. How can I say this? Here, I want to tie in with the Catholic tradition. It says that what is right is what is natural, and what is wrong is what is unnatural. God made us in certain ways—male and female, for instance— and from this, it follows that we should behave in certain ways. For instance, we should be loving and faithful to our partners. We should cherish our children. We should respect each other. Rape just doesn't fit into this picture and so is unnatural and, hence, is wrong.

My integrated position could not be working more strongly at this point. In the context of biology, someone who goes around raping is a biological pariah. Women don't want him because he is imposing himself upon them, and perhaps leaving them pregnant. In this case, they have to bear his child, no matter how biologically inferior he is and despite the fact that he is not going to be around to help with the child rearing. Men don't want him, because he is violating their wives and daughters, or those of others who are playing the social game. In the context of religion, someone who goes around raping is violating every commandment that Jesus laid upon us. Not much of the Good Samaritan here, I am afraid. Science and religion come together in an integrated whole.

Fentiman: Yes, but what about something like homosexuality? Surely that is unnatural?

Davies: Yes, indeed. You Christians are violently against homosexuality, and yet, biologically, we now know that matters are much more complex—at least, less clear-cut, than we once thought. It seems to be a paradigm case of unnaturalness, putting a penis into an anus rather than a vagina. But, any Darwinian has to take pause to think when we find how commonly humans are homosexual. There are some suggestions that up to 10 percent of men are gay, and these days the thinking is that women might not be that far behind. Some people, including Wilson, have suggested that homosexuality might be a reproductive strategy, an adaptation, put in place by natural selection. If gays help nephews and nieces—think of the popes—then this could be good biology.

Matthews: Because you are gay, it doesn't mean that you can't have kids. It's no great secret that I am a lesbian and have lived now for over ten years in a loving relationship with my partner. We have a kid, thanks to a dear and now dead gay friend who donated his sperm. I'm the biological mother and Deanna, my partner, is the social mother. In any case, until recently, I am not sure how much control women had over their reproductive fates. Straight or gay, they just got married off and had babies. But generally, you and I are on the same side. I think being gay is as natural as being straight. So, I think it is morally quite okay to be a gay man or a lesbian. Morally speaking, this is no big issue. It is not something forced. Just as much as intelligence, it is what being human is all about. Where morality comes in is in being unkind or unfaithful or some such thing.

Note that although I am going against what people thought in the past—I am not kidding myself about how lesbians used to be considered and still are in many circles—I am absolutely not excusing or finding biological justifications for any and all human behavior. I know only too well how much evil there is in the human heart. I know all about Verdun and the Somme, Auschwitz and the bombings of English and German cities, not to mention the atomic bombs over Japan. I know about Stalin's gulags and some the dreadful things of the second half of the twentieth century. For every saintly Dietrich Bonhöffer who went to the gallows for opposing Hitler, there was a Heinrich Himmler who murdered Jews as his "final solution." But, from an evolutionary perspective I expect that, too. At least, I don't expect the murder of six million Jews. I expect hatred and selfishness and more, things that get dreadfully magnified with modern technology and communications. One person can't murder six million people. But, a state geared up that way obviously can.

For me again, I have a meshing of my Christianity and biology. I think humans are torn between doing good and doing wrong. It is part of their biology. As a Christian, this is what I mean by original sin. Not someone eating an apple long ago, but the way that we humans are caught between good and ill. Jesus knew this. He hated the sin, but he did not hate us. He had mercy on us because of our fallible, finite natures. That's why he died on the cross. You might say, why bother? Now, we've got the scientific explanation, why bother with the religious explanation? Simply because I think it is true and, more than that, makes sense of things in a way that science just doesn't always capture.

Take the Germans. Go back to the beginning of the twentieth century. This was the country with the greatest culture. The universities were models for the world, especially for America. The art was at its highest form. While France had the operettas of Offenbach, the Germans had Wagner and Mahler and Strauss and the others. The country was leading the way with pensions and health care. This was the country where the Jews had been truly emancipated. There were no ghettos like those in Poland and Lithuania. And yet, this was the country that voted for Hitler. That *voted* for him! That let the thugs roam the streets smashing windows. That built up the concentration camps and killed the sick and handicapped, the Jews and the homosexuals and the gypsies and everyone else who would not go along. If the Germans had been barbarians, then we might understand Auschwitz. Not forgive it, but understand. But the Germans were not. They were the highest pinnacle of civilization. And then they sank. That's why I believe in original sin.

Fentiman: Pastor Hal has been waiting very patiently in the wings. What do you have to say, Hal?

Wallace: Thanks. I am afraid that I am going to pour a lot of cold water on what we have heard tonight. I don't deny that people on the panel are sincere. But God isn't interested in sincerity. He's interested in our doing what he wants, acknowledging him as Lord and Master. I am sure that Richard Dawkins is sincere, but he's going to hell because he does not recognize Jesus as his savior. So, let me start at the most basic. As a Southern Baptist, I deny the truth of evolution. Of course, I am not saying that there has never ever been any change. There are changes within what the Bible talks of as "kinds." I am quite happy to accept that the finches on the Galapagos came from one set of ancestors. That is just trivial change within kinds. I have said before, I certainly believe that there has been a lot of change since Noah's Flood, and it would not surprise me to learn that much change occurred within the human species. This is all microevolution. What I deny is macroevolution. We brought up this distinction in an earlier program. A horse turning into a cow. A daisy into an oak tree. And above all, a monkey turning into a man.

I don't want to say that there have been no bones found. I am sure there have. Remember, I believe that science and religion can work together. But, I just don't think that these bones mean anything very much. For all of the denial of various panel members, Genesis speaks of there having been giants before the Flood. I don't see why there shouldn't have been pygmies. Except, of course, no one is saying that these animals were actual human beings. I don't see why they shouldn't be apes that were wiped out by the Flood, just as the dinosaurs were.

Since I am sure that someone is going to bring this up, what about races? The Bible says that Ham, the son of Noah, was punished because after the Flood, he saw his father drunk and naked. Although rather mysteriously, the Bible also says that it is Ham's son Canaan who is punished. "And Noah awoke from his wine, and knew what his younger son had done unto him. And he said, Cursed be Canaan; a servant of servants shall he be unto his brethren. And he said, Blessed be the LORD God of Shem; and Canaan shall be his servant. God shall enlarge Japheth, and he shall dwell in the tents of Shem; and Canaan shall be his servant."

Rudge: My congratulations on quoting it by heart!

Wallace: Yes, I am looking forward to the day when Professor Davies can quote long passages from the *Origin of Species* by heart.

Davies: That's the whole difference between you and me. You take one book as basic that can never be changed or challenged. You know that, as a scientist, I am with Popper, who says that everything is open to debate, to falsification if the facts say otherwise. I think Darwin was a great scientist, but we have come a long way in the 150 years since the *Origin*. We discussed that last time when we talked about the fossil record before the Cambrian.

Fentiman: Okay, folks. Let's get back on topic, shall we? Pastor Hal. You were talking about race.

Wallace: Thanks. There's no doubt that the human species today is divided into different races, from the Nordic types in Europe through the Asiatic members of the species, those in Africa, and let us not forget the aborigines in Australia. The Bible is quite clear that the only humans who survived the Flood were Noah and his family, so I believe that all of the people of today are in fact descended from the sons of Noah. What we all know is that many people, especially white Americans in the South, automatically concluded that the black races, the Negroes, were the descendants of Ham, through Canaan. And they used the Bible passage that I just quoted for justification.

I want to make it quite clear that this is not the position of me and my congregation. In *Genesis Flood*, Whitcomb and Morris tackle this issue explicitly. They point out that, after the Tower of Babel, descendants of all three of Noah's sons—Japheth, Ham, and Shem—were living in western Asia. So, you really cannot tell who gave birth to which race. This point incidentally gives the lie to Professor Davies. He says that we evangelicals cannot ever change because we hold to one book as infallible. That is obviously not quite true. The Bible is infallible. But, we humans are fallible and, over time, we gain greater understanding of the meanings of the Good Book. We don't swing around recklessly. Justification by Grace is a bottom-line doctrine that will never change, but we do strive for ever-greater insight. Although I said that I know that Richard Dawkins is going to hell, I don't want to say that I think that hell is necessarily what John Calvin thought it was. Perhaps he will just enter a state of nonbeing rather than the torments of everlasting fire.

Matthews: I confess that I think that you have pseudoarguments answering a false question. You assume that race does exist and needs explaining. But the geneticist Richard Lewontin has pointed out that almost all of the variation in the human race is within groups rather than between groups. So, the idea of race is simply what is known as a social construction and, frankly, one best dropped as soon as possible.

Fentiman: I can see Martin trying to get in on this discussion, but I wonder if we could let Hal finish first, and then you can have your turn.

Wallace: There is only really one other thing I want to talk about, and again, I am sorry if I give offense. I am glad you have raised Professor Lewontin's name. He has pointed out that human sociobiology implies not just the materialist philosophy, which of course he embraces, but also what he calls "genetic determinism." What you are saying is that humans are robots—or better, marionettes—driven by the genes. The double helix, the DNA molecule, is up there behind the curtains, and we poor humans dance to its tunes. If you are bad, it is because of your genes. If you are good, it is because of your genes. You should not be condemned because you got bad genes. You should not be praised because you got good genes. Adolf Hitler and Mother Teresa are the same, except for the luck of the DNA. It's the same with things like sex. Boys rape because it is natural, it is in their genes. It's not their fault. Most of all, what is known as sexual orientation—whether you are heterosexual or homosexual—is part of biology. People lust and couple like brutes because of Darwin.

As a Christian, everything I stand for goes against this.

Davies: There you are. I have thought all along that you believe in the conflict between science and religion.

Wallace: Not at all. I believe in the conflict between false science and religion. I believe Darwinism is wrong because it implies genetic determinism. And genetic determinism is wrong because it denies free will. For the Christian, free will is absolutely at the center of religion. Adam sinned because he ate the apple. He was not compelled to eat the apple. He had the choice and he disobeyed his Lord. Jesus freely chose to die on the cross for our sins. He did not have to die for us, but he did. In turn, God asks that we freely acknowledge him and serve him for making possible eternal life. This is our choice. God has given us some directives. We are supposed to give charity to the poor. We are supposed to look after the sick. That was what my church was doing in New Orleans after Katrina. We did not do it to get into heaven. Nothing we can do can make that possible. We did it to show thanks for the gift of eternal life.

And above all, we are expected to lead our personal lives properly. That means sex. The Bible is explicit on this. Homosexual activity is condemned as wrong, as unnatural, as immoral. It could not be more clear. The Old Testament says: "If a man lies with a male as with a woman, both of them have committed an abomination: they shall be put to death: their blood is upon them." Saint Paul says: "God gave them up to degrading passions. Their women exchanged natural intercourse for unnatural, and in the same way also the men, giving up natural intercourse with women, were consumed with passion for one another. Men committed shameless acts with men and received in their own persons the due penalty for their error."

Personally, I think it is possible for people to change their sexual preferences, through prayer and support and perhaps the appropriate professional help. But, even if it isn't, that doesn't make the activity right. God expects us to behave in his ways, the natural ways, and to do otherwise is just wrong. I don't want to point the finger at anyone on this panel, but gay marriage so-called is against God's law. And that is absolute.

Matthews: Well, you are pointing the finger, of course. You are pointing it at me! I know it's Martin's turn to get into this discussion, but I just have to say something. First, if we're going to quote Leviticus, then what about:

"Thou shalt not let thy cattle gender with a diverse kind: thou shalt not sow thy field with mingled seed: neither shall a garment mingled of linen and woolen come upon thee." Anyone who said today that you shouldn't wear a coat made of linen and wool mixed would be looked upon as if they were queer in the head. You always have to interpret even the strongest statements in the Bible. In the Sermon on the Mount, Jesus is about as strongly against force as it is possible to be. But, except for pacifists like the Quakers, no one takes him literally. That was the whole point of just war theology, started by Saint Augustine around 400 CE. When is war justified and how should it be waged?

As far as sex is concerned, many of today's best biblical scholars think that the prohibitions were against ritual homosexual acts committed by heterosexuals. You often find these in native cultures. The Jews were strongly against this sort of thing, and this is what was being condemned in the Bible. Jesus never said anything on the topic and, although it is controversial, there are those who think that given his comments about the beloved disciple, he himself might have been gay. He certainly had a sympathy for the outsider, for the person who does not fit in. For myself, I just don't care if Jesus was gay or straight. It's just not relevant.

Really, though, I want to talk more about the science. There was a lot wrong with Freud, starting with his views on women. But, he did start to lead us out of the dark age on sexuality. In his *Three Essays on Sexuality*, I think first published in 1905, he argued that sexual orientation, as we now call it, what sex of a person you are attracted to, was not something that we choose freely. In fact, Freud was ahead of his time arguing that some people are gay by nature. Just as some people are gay because of their society. Freud talked about the ancient Greeks, among whom homosexual activity was the norm, but you could just as easily talk about single-sex boarding schools. And most famously, Freud argued that some people are gay because of dysfunctional families.

In the case of boys, it is because they have dominant mothers and hostile or distant fathers. Everyone—every boy, that is—has an Oedipal yearning to go to bed with mom and, in mature growth at adolescence, this gets transferred to other females. For future gay men, the link with mother is too strong to break—you dare not look at other women—so you revert to an earlier stage of development and turn to men for sexual gratification. Freud said that this is not a sickness but an immaturity and, in a famous letter he wrote in the 1930s to an American woman with a gay son, he said that her son could not be changed sexually, but that therapy might make him more comfortable and understanding of where he was at.

Obviously, I don't like the immaturity stuff—Freud, incidentally, had a parallel tale for lesbians, involving what he called a "Jocastar complex"—but I think he was dead-on about not choosing and, for all that the Reverend Hal says otherwise, not changing. The real trouble with Freud is that he had some very old-fashioned and completely wrong ideas about biology. He was an ardent evolutionist, but instead of natural selection, he believed that the main cause of change is the inheritance of acquired characteristics—the blacksmith develops strong arms and his kid is born with strong arms. This is known as Lamarckism. Freud thought that once, long ago, a band of brothers ganged up on their father and killed him, but then decided that they should not quarrel over their mother, so they instituted a taboo against sex with mothers. After the Greek figure who killed his father and married his mother, this gives rise to the Oedipus complex.

Freud thought this all actually happened and then got ingrained through a kind of Lamarckian process on the consciousness of the human race. Obviously, this can't really be so, and I am inclined to think that Freud often got things backward. As the parent of a child, I am only too aware that in our relationship, it is he who makes the running. I think children elicit behaviors from parents and the dominance-hostile emotions are reactions to the way the kid already is. In our society, fathers get cheesed off if a son says he would rather play with dolls than go to a ball game. No wonder they become distant!

But, I want to praise Freud rather than criticize him. He was emphatic, especially in his "Letter to an American Mother," that homosexuality is not a sickness and it is not a choice and it cannot be changed. And, that it is possible to be gay and to be well balanced and to lead a happy and productive life. I certainly don't think that sociobiology has all of the answers, but I look upon it as something taking the debate and our understanding one step further. We know that it's not all in the genes. Although identical twins do generally have the same sexual orientation, this is not always the case. Probably hormones have some role. It's not a question that lesbians have more testosterone than straight women. In the 1950s, they used to try these sorts of things out—giving gay men testosterone. It just made them randier gays! Modern research suggests that it is a matter of testosterone levels, for both males and females, as it affects the hypothalamus in the third to sixth months of fetal development.

I don't know if this is true or not—and I very much doubt that there is only one cause for something as complex and important as sexual orientation—but it can certainly be fitted in with biological explanations. Perhaps, for instance, under certain circumstances, mothers would be better off having sons who were gay—ready to help siblings—and so biology kicks in. Nobody is saying that there is any conscious thinking here. Natural selection has made us this way.

Davies: Let me put my two cent's worth in here. What Emily is saying sounds fantastic, but, biologically, it is far from crazy. There is a very important finding in evolutionary biology that high-status mothers tend to have sons and low-status mothers to have daughters. This follows from the fact that, in nature, females inevitably get impregnated, but males have to compete. Some have lots of offspring and some have none at all. So, if you are going to turn out kids up the totem pole, go male. Down the totem pole, go female. This is all controlled by hormones, and there is evidence that it applies to humans. So, what Emily is saying is at least plausible.

Matthews: Thanks for support from an unexpected quarter! Let me just bring what I want to say to an end. Today, we have gone beyond prescientific Christianity. We can turn to and work with science to make a unified, integrated whole. Ultimately, what we know now is that to talk of being gay as unnatural or straight as natural is not really meaningful or helpful. Apart from the fact that just about every animal species that has been studied intensely shows same-sex behavior, in humans to speak of love between men or love between women as unnatural is simply wrong. I am not saying that all actions are natural. I would think that self-mutilation is unnatural. But I am saying that sex between freely loving people is perfectly natural.

And at this point, as I said, I want to tie things in with the Christian tradition. I embrace the "natural law" position of Saint Thomas Aquinas. What is natural is good and right. God created our bodies—I think through evolution—and with that come responsibilities and duties. We should not use our bodies in the wrong way, in unnatural ways. My quarrel with the Catholic Church and with some of my very conservative Episcopalians, not to mention people like the Reverend Hal, is not over the right to be unnatural. God does not want that. It is over what comes under the cover of unnatural. I think modern science has shifted the goal lines. As one who integrates science and religion, I am happy to go with that.

Fentiman: Time to let our philosopher have his say. Martin?

Rudge: Obviously, I am happy to go right along with David Davies and his lucid description of human evolution. I don't suppose that anyone thinks we have the last answers, but clearly, we now have a pretty good idea about the paths and the causes of human evolution. I should say that I am also very impressed with Emily Matthews's grasp of human science, both about Freud and about human sociobiology. Much as I enjoy reading Freud—and I do think he had some really deep insights—I confess that I do wonder at times whether it isn't really all a big fairy story. I confess, also, that I often feel the same way about human sociobiology, although, more and more, when it comes to social behavior and ethics, I think there is something really significant going on.

Of course, if you do think that biology can speak to morality, then particularly if you are thinking about religion, you must grasp the nettle of determinism. You folks have been talking about genetic determinism. I think an awful lot of hot air has been spilled on this one. If you accept science and include humans, then at one level, you are really bound to be deterministic, for all of the qualifications you might want to make about quantum mechanics. Modern science presupposes that the world is governed by unbroken law. So, in a sense we have to say that humans—as part of the world—are bound by laws and part of the causal process. And, obviously, we are in some sense. Firms and politicians pour a lot of money into advertising because they know that humans are predictable.

But, I am not sure that this is the end of discussion. David Hume, to talk again about my hero, said that unless we were determined in some sense, we couldn't even have free will! He said that the real distinction is not between free will and determinism, but between free will and constraint. If there are no laws, nothing is determined; then, you do not have freedom. You have chaos. If I pull out a gun and kill you all, I deserve to be punished because I am a moral agent—I have reason and training and whatever. Things in the past that are supposed to guide and control me now. If, however, I do it because the laws of nature have suddenly broken down, then why blame me? It just happened. For Hume, what counts is whether I did the act because I could or because I was forced to do so. If someone had hypnotized me on the way to the studio, then I am no longer a free agent and should not be condemned—even though, both free and controlled, I am part of the causal network.

Wallace: I know all about this. Your position is called compatiblism, because you think freedom and laws can go together. Not everyone, including me, agrees with this. We think that humans are autonomous agents and that, in an important sense, we can transcend the causal order. Human rationality makes for a different dimension from simple causal chains.

I should say, incidentally, that talking of human rationality makes me realize that through this whole series, we have been avoiding the elephant in the room. What about human rationality? What about consciousness? How can you have a nonreligious explanation for this most vital phenomenon of all? Mechanistic science says nothing about it. Mechanistic science can say nothing about it. "And the Lord God formed man *of* the dust of the ground, and breathed into his nostrils the breath of life; and man became a living soul." That is all you can or need to say on the subject.

Fentiman: Hold on the consciousness issue for just one moment. Martin, finish what you were saying about free will and determinism.

Rudge: Hal is absolutely right to point out that mine is only one position, and we could argue the merits of the positions until the cows come home. The point I really want to make is that I don't think that science, as such, forces you into one crude stand on free will. You can have free will and science. Claims about genetic determinism are a bit stronger. These are claims that everything we humans do and think are simply functions of the genes. You are clever, thank the genes. You are stupid, blame the genes. No one thinks that only the genes are important. I speak English rather than French because I was born in Baltimore rather than Paris. No one thinks the genes unimportant. I speak a language rather than communicate through chemicals because I have a human gene complement rather than that of an ant.

In fact, drawing attention to the ants is a good way to make an important point about genetic determinism. Obviously humans are not ants. In what way? The thing is that ants are obviously really determined—preprogrammed, as one might say. They don't think or decide for themselves. They do what they do because their genes tell them to. Humans on the other hand have dimensions of freedom. Should I take that last piece of cake or will I look like a greedy pig? Should I do my homework or should I watch TV and hope to get someone else's notes tomorrow before class? Both ants and humans are bound by law, both ants and humans are influenced by genes, but humans have something that the ants don't have. A capacity for deciding and changing course. A good analogy is between simple rockets that are fired at a target, and more sophisticated rockets that can change course in midair as the target moves. Ants are cheap rockets. We cost more. No wonder mother ant has millions of offspring and human parents have but a few!

I agree that this still does not solve the problem of how much the genes influence human behavior. Someone like Dick Lewontin wants to be closer to the culture end of things. Someone like Ed Wilson wants to be closer to the biology end of things. Emily has pointed out that there are no simple answers. Whether you are gay or whether you are straight is not solely a function of the genes, because sometimes identical twins go different ways. But generally, they don't, and this suggests some genetic component—an important genetic component. It is the same with other things. Growing up in a middle-class family is going to improve your intelligence. But it does not determine everything. Adopted children still swing back in the directions of their birth parents.

Of course, a lot of this is bound up with race issues. Once you start saying that the genes really count, then you are in danger of blaming blacks alone for their poverty rather than their conditions, to which we all contribute. We know that things cannot possibly be as simple as that. When eastern European Jews flooded into America at the beginning of the last century, intelligence tests showed that, on average, they were pretty subnormal. I don't think that anyone today would want to say that American Jews are subnormal. Dave, I checked your curriculum vitae before we started this series and even I am impressed! Harvard, Rhodes Scholar, Guggenheim Fellowship.

Davies: I am a paradigm case of what you are talking about. My great-grandfather was an immigrant from Lithuania. My grandfather worked in a cut-price shoe store all of his life, while my grandmother worked in a sweatshop. Both of them spoke and thought in Yiddish to the end of their days. Both of them thought education was the be-all and end-all, and they were right! My dad was a doctor—a hell of an achievement for a Jew in those days, since all of the prestigious colleges like Harvard had strict quotas on Jews—and I am the fourth generation. I was expected to work, work, work, and I did! I wish my kids felt the same way. One is majoring in philosophy, for goodness' sake!

Matthews: Cheer up! He might have found religion and become an Episcopalian minister!

Fentiman: Time is moving along. Finish up, Martin.

Rudge: One way, a way we've heard talked about this evening, to avoid the race issue is to deny that there is genuinely a notion of race. Jews cannot be brighter or thicker than the rest of us if there is no Jewish race—no race, that is, in the biological sense. This was Lewontin's argument. There is more variation within groups than between them. But recently, that argument has been shown fallacious. What Lewontin said is true, but various features get clustered in the groups—various genes, ultimately. So, you can separate out human races, even though today they are clearly breaking down thanks to travel and so forth. Again, we have the example of the Jews in America. So many today marry Gentiles that in a hundred years, you will be pushed to find anyone who is of pure Jewish descent or anyone who does not have some Jewish blood in their veins.

More seriously, you can't get around the race issue by denying its existence. The fact is that if you take biology seriously, then you have to allow the possibility of different behavioral and intellectual traits, as well as physical features. Ed Wilson argues that Orientals have trouble synthesizing alcohol—they get drunk faster—and he thinks that this cascades up into differences in social behavior. I am not sure about that, but it is plausible. What I would say before we dash out and say that blacks are not as bright as whites, we should level the playing field. Growing up in the slums of Chicago's South Side, with a single mother on welfare, is not exactly conducive to ending up at a snobby college like Northwestern in the middle-class suburbs like Evanston, up the lake.

Fentiman: We are coming to the end of our time. But before we finish, I do think we should say something about the topic that the Reverend Hal raised. What about consciousness? I am not sure that it is the only issue, or even the most important issue, on the science-religion interface. But it is an issue. Rather than arguing, I wonder if I could get a statement from each of you on the subject. I think it would be really good if you would try to relate your answers to the four positions you claim to take on the science-religion question: conflict; integration; dialogue or working together; and independence. That way we can remind listeners of your basic thinking, and illustrate how it works in action. Start with David Davies.

Davies: My position is simple. Consciousness is a perfectly natural thing, something that requires no fancy explanation. It is obviously something that is brought about by natural selection, and it is clearly adaptive. Organisms that can think can puzzle out issues in ways that organisms that cannot think are unable to do. I don't mean that consciousness is always the best route to take. Consciousness is obviously linked to big brains and, as we've heard, big brains are expensive to maintain. So, when meat is not readily available, you might be better off going dumb and cheap! How exactly consciousness works is something that is still a bit of an open question. We know that various parts of the brain do various things that manifest themselves in consciousness. Speech abilities, for instance, are connected to Broca's and Wernicke's areas. Stephen Pinker suggests that one major function of consciousness is that of a filter, a guide and coordinator to all of the information thrown up by the brain: "Information must be *routed*. Information that is always irrelevant to a kind of computation should be permanently sealed off from it. Information that is sometimes relevant and sometimes irrelevant should be accessible to a computation when it is relevant, insofar as that can be predicted in advance."

The main thing is that you don't need any of this breath-of-life nonsense. God is out. The way I like to think of consciousness and its relationship to the brain is that of the hardware and the software of a computer. The brain is the hardware. The mind, consciousness, is the software. The brain is all of the physical electrical circuits. The mind is the program—Word or Procite or PowerPoint. Or a movie that you have on a DVD. The point is that we can now see that there is nothing beyond the material or physical. So, the clash between science and religion is absolute.

Fentiman: Now, on to you, Emily.

Matthews: As an integrationist, I agree with David that we have got to get away from old-fashioned miracles—breathing in life—although I don't mind it as a metaphor. But as an evolutionist, I see mind and brain co-evolving, different sides of the same thing. I believe my position is known by philosophers as an "identity theory," and that it goes back at least to the seventeenth-century Dutch Jewish philosopher Spinoza. I think mind and brain are one, but that they are different perspectives. Take an analogy. You might know a church member who is a bit of a grump, who really is not very sociable at all. You might also know that in your community there is someone who has given a large anonymous gift to the church to support the outreach program to the poor and homeless. These are two completely different people. One you don't much like. One you admire immensely, especially because they are doing as Jesus asked and making their charity in private, not asking for reward. Except, later you find out that these are one and the same person. It's the same with the mind and the body, and I think that both as a Christian and as someone interested in psychology, I have a perfectly integrated picture.

You see, although at one level, I want to go along with David Davies and not appeal to miracles or anything that does not agree with modern science, at another level, I find his thinking inadequate. His analogy—or perhaps he thinks it no analogy, but literal—about the computer does not work. Agreed you have hardware and software. But what is the software? Is it a physical disk with certain things etched on to it or is it the information that it carries? If it's the information—let's say a James Bond movie—then it seems to me you have still got to go from the physical etchings to the beautiful girl seducing Bond and leading to his woes at the hands of the master criminal.

Davies: I have a feeling that you know more about these movies than I do!

Matthews: Whether I do or I don't, I still think there is a gap in your story. The physical disk is purely material, but the information that it carries is not. It does not seem to me to be absolutely impossible for a creature from outer space to have a completely different language and a computer that translates the disk into something else. A "chick flick," as they are called. Suppose I have the numbers 1234. In my language this translates as S O U P—soup. In its language—same disk—this translates as D I R T—dirt. So, I don't think your position works. Note that I am not denying any of your science. I am simply denying the philosophical consequences you want to draw from it.

I should say that this for me is the real miracle. We should stop looking for rare conjuring tricks and start appreciating the things under our noses. I'm not looking for breaks in the natural order of things. I think the very fact of consciousness is a miracle. That dead matter somehow can give rise to thinking is a miracle, every time and all of the time. My God is really terrific!!

Rudge: Your position is known as "natural supernaturalism." It was endorsed by a Scottish writer of the Victorian era, Thomas Carlyle. Nobody much reads him these days because his style is so convoluted, and he did end with some vile racial attitudes. But in an early work, *Sartor Resartus*, he argued that miracles are all a matter of perspective. "What specially is a Miracle? To that Dutch king of Siam, an icicle had been a miracle." He went on to say: "Custom"—and I am paraphrasing a bit here—"doth make dotards of us all. Innumerable are the illusions and legerdemain-tricks of Custom: but of all these, perhaps the cleverest is her knack of persuading us that the Miraculous, by simple repetition, ceases to be Miraculous." Carlyle would agree that consciousness is miraculous, but because it is so commonplace, we fail to appreciate it for what it is worth. He sneered at Dr. Johnson, the lexicographer, for his search for ghosts. We are all ghosts, inasmuch as we are conscious beings within a physical frame.

Fentiman: Another Rudgian footnote! Emily has had her turn. Now, Pastor Wallace.

Davies: No, just a sec. Since Emily's criticizing me on my grounds, let me criticize her on her grounds. Suppose we accept what you say. Then what about all of your beliefs about life after death? What about the Jesus claim that we can have eternal salvation? If the mind and the brain are one, then when one goes, so goes the other. If the brain decays after death, and it certainly does, then there can be no mind. At least on my position, God can have a kind of megacomputer where everything is stored. On your position, the end of life is the end of existence.

Matthews: Actually, not quite. The belief that body and mind are separate is a Greek belief. The Jews thought that it is a package deal, and this is the position of Saint Paul. He says that there will be a resurrection of the body—a spiritual body, admittedly, but a body nevertheless. So with my spiritual resurrection, I fully expect to have a mind. Spiritual also, but what else would you expect in the kingdom of heaven?

Fentiman: Now, Harold Wallace, please.

Wallace: Well, of course I do agree with Emily about the spiritual resurrection of the body, but, remember that I think that sometime soon those still living will see the return of the Lord. After the chosen have been raptured up. It seems to me that the spiritual body and the physical body might not be so far apart. Also, I want to talk in terms of the soul as much as the mind. I am against abortion—all abortions, at any time, for any reason. I am also against euthanasia—all euthanasia, at any time, for any reason. We all know of the case of the woman in Florida, Terry Schiavo, who was in a coma or something like that, and who was starved to death because her husband ordered it. I say that this was murder. The point I want to make is that I don't necessarily equate soul with consciousness. I don't care if Terry was conscious. She was still someone with an immortal soul. The same is true of a fertilized egg from the moment of conception. It has a soul, and killing it is murder.

So, I am not impressed with Professor Davies, who thinks the mind is just a CD or a DVD. Nor am I impressed with the Reverend Matthews who thinks that only if the brain is working is the mind working. I think the soul is bound up with consciousness—perhaps without a soul you cannot have full consciousness—but I think the soul is more than this. For this reason, I have to have a miracle to start human life—a miracle each time that human life starts—and this is something beyond science. I appreciate the sentiment that the very fact of consciousness is a miracle, but to me the real power of God is his constantly making and renewing his creation.

Although Emily Matthews is not that keen on Greek philosophy, I do think that perhaps body and mind, body and soul, are separate things. In this degree, I am a dualist. I know that dualism isn't very popular with philosophers because they cannot see how body and mind work together. But I would say that they have brought this problem on themselves by tying their thinking to naturalism. If they could see that science does not have all the answers, they could see that a God who knows when the meanest sparrow falls, also can keep body and mind, body and soul, working together. It is as simple as that.

Davies: I don't see much dialogue in your thinking. How do science and religion work together?

Wallace: I think that you yourself have given the answer, even though you did not intend to. As a scientist, you have shown that science has no answer. It is like irreducible complexity. There comes a point when you simply have to appeal to an Intelligent Designer. In any case, we are all conscious here. We all know that our thinking is not just bricks and stone. Sometimes, you need to stay with the obvious, rather than get caught up with fancy analogies about computers.

Rudge: My turn now, I think. In a way, I am going to be a bit of a disappointment among all of these confident people who know how to resolve the body-mind problem. I think Emily is quite right about the problems with David's information-bearing CDs. The philosopher John Searle has made a similar point with a story about a Chinese box. You might feed in Chinese questions on the one side and—thanks to someone inside, with a rule book, who matches symbols of one kind with symbols of another kind—get out Chinese answers on the other. This does not mean that the person inside knows what on earth he is up to. He is just following mechanical rules rather than thinking about the problem. He is not thinking in Chinese, even though to the outside, it may seem that he is. Information, thinking, is more than the physical or mechanical manipulation of the world.

I confess, however, that I have problems with Emily's position also. I just don't see an identity between brain and mind. Your grumpy parishioner and your generous benefactor are the same kind of thing—they're both people, even though you didn't realize that they are one and the same person. Brain and mind are different kinds of things—that is why Hal's dualism seems so obviously right, even though it does have all of the horrendous problems about why the two should have anything at all to do with each other, let alone coordinate in the way that they clearly do. Hal, like earlier dualists—most notably Descartes—has God doing the work for them, but that seems to me to be a violation of the spirit of science.

So, if I say a plague on all of you, what am I left with? One thing I am fairly certain of is that we are not going to solve things with our present way of looking at things and of doing science. I don't mean that we are not going to learn a lot more by looking at brains and seeing how they relate to mental states. We have learned a huge amount and I suspect we will learn more. I am also open to finding out more about the selective virtues—or limits—of minds. At that level, I don't want to deny anything that Dave has said. But the ultimate question about the nature of mind seems to me to be beyond present science, in principle. I don't want to say that it is a science stopper, because I am open to other ways of solving the issues. But not our ways.

Why do I say this? Well, this is a consequence of my basic theory of science and religion that I have been trying to get out since the first program. I started to expound my thinking in, I think, the second program, when I talked about metaphor and science. I agree with Emily that the big change in the sixteenth century was the change from the organic metaphor—the world as an organism—to the mechanistic metaphor—the world as a machine, a clock. The world, including us humans, is regarded as one big machine, running according to fixed laws of nature. I think that Copernicus and his successors established the metaphor in the physical world. Darwin established the metaphor in the biological world. And perhaps Freud and the social scientists established the metaphor in the human world.

It is a really powerful metaphor. Unlike Emily, I don't regret the end of the organic metaphor. I don't see much mileage in the Gaia hypothesis, and I certainly don't see the metaphor as being pure or good in a way that the clock metaphor is not. It is a bit of a low blow, but the keenest boosters of the organic metaphor in the twentieth century were the National Socialists. They loved the idea of the organic state and of Jews being parasites that preyed upon it. But, while the mechanistic metaphor has great virtues—it leads to modern science—like all metaphors, it has limits. Some questions are simply not asked nor can they really be asked. If I say that my love is like a red, red rose, you might infer that she is beautiful. I suppose that you might even infer that she has sunburn! But, you are not saying anything about her abilities in mathematics.

Similarly with the clock metaphor. With machines, you simply have to take their materials as given in some ultimate sense. You cannot ask about first origins. Likewise, although you can use machines for good and ill, in themselves they have no moral value. Of course, this ties in with everything

I have been saying about science and values. Pertinently here, I don't think that the clock metaphor speaks to consciousness. Machines don't think! They don't think, even in principle. Of course, there are computers doing calculations and so forth. Some clever programs can even make you think you are having a conversation. But ultimately, they are just bits of clockwork—or more complex pieces, including electronic pieces—going through the motions. And, if you have a seeming counterexample like the computer Hal in the movie *2001*, then ultimately you ask yourself: Is this just a machine or has it got to the stage where it has moved beyond machine and become conscious? Personally, I think our Hal is a thinking being and the computer Hal was not, but the choice is there and must be made one way or the other. Where does this all lead? I don't think our science even gets to first base with the problem of consciousness.

Wallace: It seems to me that, finally, for all of your scorn for thinking like mine, you are embracing "science stoppers" just as much as I am. You are saying science cannot solve the problems of consciousness. That's my position, but at least I don't conceal it. It may seem rude to accuse you of hypocrisy, but I don't see why the charge shouldn't stick.

Rudge: Fair comment, and I do worry about this. But I think there is a difference. You are saying some problems cannot be solved, period. At least, there is no scientific solution, and we must turn to an Intelligent Designer. The flagellum of the bacterium, for instance. I'm not quite saying that. I'm saying that with the present models or metaphors underlying science, I don't think we can solve the problem. I'm leaving it open for someone to come up with a new metaphor for science that will do the job. What is this model or metaphor? Well, if I could tell you that, I would be the new Copernicus or Darwin!

Although, I do have to reveal my hand a little more than that. A number of philosophers now suggest that perhaps the human thinking apparatus is just not strong enough to solve the problem of consciousness. As a Darwinian, I am inclined to buy into that line of thinking. I don't think adaptations for getting out of the jungle and onto the plains are necessarily going to be good enough to solve all of the big questions of the universe. For me, the wonder is that we get as far as we do. Richard Dawkins has admitted this much. He says: "Modern physics teaches us that there is more to truth than meets the eye or than meets the all too limited human mind, evolved, as it was, to cope with medium-sized objects moving at medium speeds through medium distances in Africa." I don't mean that there is no solution of a natural kind, which seems to me to be your position, Hal, but that we cannot get to the solution. But I doubt we can ever have a proof that we cannot get to the solution, so we have to keep trying.

This is where my belief that science and religion properly understood are systems apart really comes into play and shows its strength. I think that there is a list of things that a science based on the mechanical model or metaphor simply rules out of discussion. As you will have gathered from the things I have been saying over the weeks, these include ultimate origins, morality—at least, in the sense of some ultimate justification—consciousness. To this list, I would perhaps add questions about the meaning of it all. I am a skeptic about all of these. I don't have answers. But I think it is possible and legitimate for a religious person to step in here and say that he or she wants to fill the gaps with religion. God is responsible for ultimate origins, God is the being that set the rules of morality, God knows about consciousness, and God gives the ultimate meaning to existence. Science is talking about one dimension of life and religion is talking about another. The two are in harmony, but they do not overlap or take on the roles of the other. Independence, in other words!

Fentiman: And with that, I am afraid, we must draw things to an end. We have run out of time. Our five-part series is finished. Those of you at home who want to return to the issues and who would like transcripts of these programs will be pleased to learn that the series will be edited and published next year. Look out for details. For now, it is my pleasure to thank David Davies, Martin Rudge, Emily Matthews, and last but certainly not least, Harold Wallace. I am your host, Redvers Fentiman, and I wish you all a very good night!

~

Endnotes

page:line

3:1 This quote is from a 1999 PBS TV program, *Faith and Reason*.

3:24 Karl Popper's classic work on the nature of science is *The Logic of Scientific Discovery*, London: Hutchinson, 1959.

4:13 An excellent collection on the relationship between Christianity and science is edited by David C. Lindberg and Ronald L. Numbers, *God and Nature: Historical Essays on the Encounter between Christianity and Science*, Berkeley: University of California Press, 1986. A very important new interpretation of the meaning of the Protestant Reformation for science is by Peter Harrison, *The Bible, Protestantism, and the Rise of Natural Science*, Cambridge: Cambridge University Press, 2006.

4:19 The classic work on the Copernican Revolution is still Thomas S. Kuhn, *The Copernican Revolution*, Cambridge, MA: Harvard University Press, 1957.

5:18 This comment was made on the *700 Club* by Pat Robertson in 2005, after a federal judge in Dover, Pennsylvania-ruled that Intelligent Design Theory has no place in the biology classes of state-supported schools.

5:25 Ian Barbour, *Religion and Science: Historical and Contemporary Issues*, San Francisco: Harper, 1997.

6:4 Alfred North Whitehead, *Process and Reality: An Essay in Cosmology*, New York: Free Press, 1978. These were the Gifford Lectures, originally given in Edinburgh in 1927–1928.

6:14 This is from a 2006 BBC television program called *The Root of All Evil*. Richard Dawkins's most systematic attack on religion is his book, *The God Delusion*, New York: Houghton Mifflin, 2006.

6:29   This is from a review of the *Origin of Species*, written in 1860 by Thomas Henry Huxley, and reprinted in his collection of essays, *Darwiniana*, London: Macmillan, 1893.

8:22   This controversial charge is made by Daniel Goldhagen, *Hitler's Willing Executioners: Ordinary Germans and the Holocaust*, New York: Alfred A. Knopf, 1996. A vigorous response is Norman Finkelstein and Ruth Birn, *A Nation On Trial: The Goldhagen Thesis and Historical Truth*, New York: Henry Holt, 1998.

10:13   The connection—real or apparent—between Darwinism and National Socialism is a much-debated issue. For opposing views, see D. Gasman, *The Scientific Origins of National Socialism: Social Darwinism in Ernst Haeckel and the Monist League*, New York: Elsevier, 1971, and A. Kelley, *The Descent of Darwin: The Popularization of Darwin in Germany, 1860–1914*, Chapel Hill: University of North Carolina Press, 1981. For a really bloodthirsty application of Social Darwinism, see F. Von Bernhadi, *Germany and the Next War*, London: Edward Arnold, 1912. In *Mein Kampf*, Hitler does actually say: "He who wants to live must fight, and he who does not want to fight in this world where eternal struggle is the law of life has no right to exist."

10:27   A good discussion of the ideological roots of National Socialism can be found in S. Friedlander, *Nazi Germany and the Jews: The Years of Persecution 1933–1939*, London: Weidenfeld and Nicolson, 1997. There has been a great deal written on Social Darwinism. R. Bannister, *Social Darwinism: Science and Myth in Anglo-American Social Thought*, Philadelphia: Temple University Press, 1979, is a good place to start. Kropotkin's classic work, *Mutual Aid*, has been reprinted many times. Background material, tying his thinking in with biology, is Dan Todes, *Darwin Without Malthus: The Struggle for Existence in Russian Evolutionary Thought*, New York: Oxford University Press, 1989.

11:10   Phillip Johnson, *Darwin on Trial*, Washington, DC: Regnery Gateway, 1991.

12:18   A. Plantinga, "When Faith and Reason Clash: Evolution and the Bible," *Christian Scholars Review*, 1991, 21: 8–32. Reprinted in D. Hull and M. Ruse, eds., *Readings in the Philosophy of Biology*, Oxford: Oxford University Press, 1998, 674–697.

12:35   William Whewell, *The History of the Inductive Sciences*, London: Parker, 1837, 3: 588.

14:11   Saint Augustine's claim that God stands outside time can be found in his *Confessions*.

14:19   P. Hefner, *The Human Factor: Evolution, Culture, and Religion*, Minneapolis: Fortress Press, 1993.

15:27   C. Merchant, *The Death of Nature: Women, Ecology, and the Scientific Revolution*, San Francisco: Harper, 1990.

17:1   James Lovelock, *Gaia: A New Look at Life on Earth*, New York: Oxford University Press, 2000.

17:27   Charles Hartshorne, *Man's Vision of God and the Logic of Theism*, New York: Harper, 1941.

19:18   Of the many books available on the history of American Christianity, Mark Noll's, *America's God: From Jonathan Edwards to Abraham Lincoln*, New York:

Oxford University Press, 2002, is simply superb, especially for an understanding of how biblical literalism became such a force in the land.

19:39 Ibid., 11.[AQz2]

20:31 Ursula Goodenough, *The Sacred Depths of Nature*, New York: Oxford University Press, 1998.

21:2 Jan Smuts, *Holism and Evolution*, New York: Macmillan, 1926.

21:13 Stephen Jay Gould, *Rocks of Ages*, New York: Ballantine Books, 1999.

21:24 I discuss some of these issues more formally in my *Can a Darwinian be a Christian? The Relationship between Religion and Science*, Cambridge: Cambridge University Press, 2001.

23:24 The classic work by an American on this topic is Langdon Gilkey, *Maker of Heaven and Earth: The Christian Doctrine of Creation in the Light of Modern Knowledge*, New York: Anchor, 1959.

28:18 There are lots of books on Darwin and his theory, and on the modern versions. A couple of my books might help. *The Darwinian Revolution: Science Red in Tooth and Claw*, 2nd ed., Chicago: University of Chicago Press, 1999, covers a lot of the history. You will find full details of points made by the discussants. *Darwinism and Its Discontents*, Cambridge: University of Cambridge Press, 2006, talks about the modern theory, its triumphs and its critics. It gives full details of the significance of genetics for evolutionary thinking.

29:25 Normally, when a character refers to a classic work, I do not bother to give a reference to a particular edition. Most are as good as the others. The case of Darwin's *Origin* is a little bit different. The work went through six editions, with Darwin adding and revising very considerably. It used to be the case that most reprints were of the final, sixth edition of 1872. However, scholars today much prefer the first edition of 1859, before Darwin started chopping and changing, often for reasons that we find less than compelling. For instance, the physicists claimed that there was too little time for a leisurely mechanism like natural selection to have full effect. Darwin responded by supplementing selection with additional mechanisms, like Lamarckism, the inheritance of acquired characteristics. Now, we know that this is a false mechanism speaking to a pseudoproblem, for with the discovery of radioactive decay and its heating effects, the age of the earth has been much extended.

There is a facsimile edition of the first edition of the *Origin* (published by John Murray in London), edited by the well-known evolutionist Ernst Mayr, and published by Harvard University Press, in Cambridge, Massachusetts.

29:34 Ibid., 63–64.

30:8 Ibid., 80–81.

32:10 William Whewell, *The Philosophy of the Inductive Sciences*, London: Parker, 1840.

33:16 A lot of people wish that Darwin and Mendel had got together back in the 1860s, and done then what eventually only really came to fruition in the 1930s. The fact is that neither scientist was really thinking in such a way that he could have appreciated the work of the other. At the beginning of the twentieth century, selectionists did not think that Mendelism was very significant, and Mendelians returned the compliment. Mendel read the *Origin* (in German translation), but his annotations were all about the religious implications! He was thinking as a monk, not a geneticist. Darwin did not read Mendel, but he had works that referred to Mendel and he could have found the papers had he been searching. There is just a slight chance that Darwin and Mendel were present at the same scientific congress in London in the early 1860s, but no reason to think that they were introduced.

33:7 The point about the significance of adaptation in modern evolutionary thinking cannot be overemphasized. My *Darwin and Design: Does Evolution Have a Purpose?* Cambridge, MA: Harvard University Press, deals with this topic, both historically and conceptually.

34:28 Dawkins makes this claim at the beginning of his *The Blind Watchmaker*, New York: Norton, 1986.

34:36 The original paper on punctuated equilibrium is Niles Eldredge and Stephen Jay Gould, "Punctuated Equilibria: An Alternative to Phyletic Gradualism," in T. J. M. Schopf, ed., *Models in Paleobiology*, San Francisco: Freeman, Cooper, 1972, 82–115. The theory is discussed in lots of places, including my *Darwinism and its Discontents* and also in my *Mystery of Mysteries: Is Evolution a Social Construction?* Cambridge, MA: Harvard University Press, 1999.

35:3 The Arkansas Creation Trial (1981) and the more recent trial in Dover, Pennsylvania (2005), both over the admissibility of religious perspectives on life's history into state-supported biology classes are covered in Robert Pennock and Michael Ruse, eds., *But Is It Science? The Philosophical Question in the Creation-Evolution Controversy*, 2nd ed., Buffalo, NY: Prometheus, 2007.

35:30 Stephen Jay Gould and Richard C. Lewontin, "The Spandrels of San Marco and the Panglossian Paradigm: A Critique of the Adaptationist Program," *Proceedings of the Royal Society of London, Series B: Biological Sciences*, 1979, 205:581–598.

36:27 I talk about this in *Darwin and Design*. The classic work on the topic is E. S. Russell, *Form and Function: A Contribution to the History of Animal Morphology*, London: John Murray, 1916. It has been reprinted in paperback by the University of Chicago Press.

37:16 Pastor Hal's thinking reflects the "Creation science" of John C. Whitcomb and Henry M. Morris, *Genesis Flood: The Biblical Record and Its Scientific Implications*, Philadelphia: Presbyterian and Reformed Publishing Company, 1961. The definitive history of the Creationist movement is Ronald L. Numbers, *The Creationists*, 2nd ed., Cambridge, MA: Harvard University Press, 2006.

38:12 What most evangelicals tend to gloss over is the part of the prophecy that most of the returning Jews will be killed and those remaining will convert to Christianity.

38:35 "When Faith and Reason Clash," 685.

39:2 S. Wavell and W. Iredale, "Sorry Says Atheist-in-Chief: I Do Believe in God after All," *Sunday Times* (London), December 12, 2004, 1, 7.

39:6 Leslie Orgel, "The Origin of Life: A Review of Facts and Speculations," *Trends in Biochemical Sciences*, 1998, 23:491–500.

40:3 Gould says this in Stephen Jay Gould and Niles Eldredge, "Punctuated Equilibria: The Tempo and Mode of Evolution Reconsidered," *Paleobiology*, 1997, 3:115–151.

40:8 Many people make this claim, including Phillip Johnson in *Darwin on Trial*.

40:13 And Karl Popper, *Unended Quest: An Intellectual Autobiography*, LaSalle, IL: Open Court, 1976.

40:24 *Darwin's Black Box: The Biochemical Challenge to Evolution*, New York: Free Press, 1996.

41:22 See Ernan McMullin's introduction to his edited volume, *Evolution and Creation*, Notre Dame, IN: University of Notre Dame Press, 1985.

42:16 A revisionist account of the Scopes Monkey Trial is E. Larson, *Summer for the Gods: The Scopes Trial and America's Continuing Debate over Science and Religion*, New York: Basic Books, 1997.

43:2 There is considerable debate about whether one can properly call Intelligent Design Theory a form of Creationism. Most ID supporters are not biblical literalists and Behe, a Roman Catholic, accepts a great deal of evolution and had little time for the eschatological speculations of evangelicals. On the grounds that they do share a moral agenda, I defend the link in my *The Evolution-Creation Struggle*, Cambridge, MA: Harvard University Press, 2005.

43:13 This example comes in *The Blind Watchmaker*.

45:12 This is from a letter that he wrote, reprinted in C. S. Dessain and T. Gornall, eds., *The Letters and Diaries of John Henry Newman*, XXV, Oxford: Clarendon Press, 1973, 97.

46:16 Sewall Wright, "The Roles of Mutation, Inbreeding, Crossbreeding and Selection in Evolution," *Proceedings of the Sixth International Congress of Genetics*, 1932, 1:356–366.

47:6 Thomas Kuhn, *The Structure of Scientific Revolutions*, Chicago: University of Chicago Press, 1962.

47:30 The best scientific refutation of Intelligent Design Theory is by Kenneth Miller, *Finding Darwin's God*, New York: Harper and Row, 1999.

48:16 R. F. Doolittle, "A Delicate Balance," *Boston Review*, 1997, 22 (1): 28–29.

48:30 E. Meléndez-Hevia, T. G. Waddell, and M. Cascante, "The Puzzle of the Krebs Citric Acid Cycle: Assembling the Pieces of Chemically Feasible Reactions, and Opportunism in the Design of Metabolic Pathways during Evolution," *Journal of Molecular Evolution*, 1996, 43:293–303.

50:27 I am covering ground discussed in detail in my *Mystery of Mysteries*.

54:23 For much more on the origin-of-life question, look at my *Darwinism and Its Discontents*.

55:13 The Paul Davies quote is from *The Search for the Origin and Meaning of Life*, New York: Simon and Schuster, 1997, 20.

56:9 Henri Bergson, *Creative Evolution*, New York: Holt, 1911.

56:30 The key work by Friedrich Engels is *The Dialectics of Nature*. It was written around 1880, and not really finished. It did not appear in English until between the wars. Somewhat typically, Haldane's piece appeared in a popular journal (he was a great writer for the masses), "The Origin of Life," *The Rationalist*

Annual, 1929:1–10. Oparin's first piece appeared a few years earlier, albeit in Russian. A translation is given in J. D. Bernal, ed., *The Origin of Life*, Cleveland: World, 1967: 199–234.

57:23 Evo-devo is an incredibly exciting new area of evolutionary biology. A terrific popular introduction is Sean B. Carroll, *Endless Forms Most Beautiful: The New Science of Evo Devo*, New York: Norton, 2005.

59:17 Karl Popper, *Objective Knowledge*, Oxford: Oxford University Press, 1972.

60:15 Stuart Kauffman, *The Origins of Order: Self-Organization and Selection in Evolution*, Oxford: Oxford University Press, 1993; D. W. Thompson, *On Growth and Form*, Cambridge: Cambridge University Press, 1917; Brian Goodwin, *How the Leopard Changed Its Spots*, Princeton: Princeton University Press, 2001.

61:29 S. F. Gilbert, J. M. Opitz, and R. A. Raff, "Resynthesizing Evolutionary and Developmental Biology," *Developmental Biology*, 1996:173, 357–372, 368.

62:13 I discuss in detail the problem of evil in the light of evolutionary biology in *Can a Darwinian be a Christian?*

63:25 1 Timothy 2:11–12; Galatians 3:28.

64:14 Letter to Asa Gray, May 22, 1860. It is reprinted in *The Complete Correspondence of Charles Darwin*, Cambridge: Cambridge University Press, 1985, 8:223–224.

64:26 G. C. Williams, "Mother Nature Is a Wicked Old Witch," in *Evolutionary Ethics*, ed., M. H. and D. V. Nitecki, Albany: State University of New York Press, 1993.

64:27 Richard Dawkins, *River Out of Eden*, New York: Basic Books, 1995, 133.

65:7 Keith Ward, *God, Chance and Necessity*, Oxford: Oneworld, 1996.

67:10 T. S. Ray, "An Approach to the Synthesis of Life," in *The Philosophy of Artificial Life*, ed. M. Boden, Oxford: Oxford University Press, 1996, 111–145.

69:33 Richard Dawkins, "Universal Darwinism," in *Molecules to Men*, ed. D.S. Bendall, Cambridge: Cambridge University Press, 1983, 423.

70:23 The classic work on the anthropic principle is J. D. Barrow and F. J. Tipler, *The Anthropic Cosmological Principle*, Oxford: Oxford University Press, 1986.

71:6 Alfred Russel Wallace, *Man's Place in the Universe*, London: Chapman and Hall, 1903, 256–257.

73:8 Steven Weinberg, "A Designer Universe," *New York Review of Books*, 1999, 46 (16): 46–48.

78:24 The fascinating case of the trilobite eyes is discussed in E. N. K. Clarkson and R. Levi-Setti, "Trilobite Eyes and the Optics of Descartes and Huygens," *Nature*, 1975, 254:663–667.

78:29 Stephen Jay Gould, *Wonderful Life*, New York: Norton, 1989.

79:27 *Genesis Flood*, 275.

80:24 I talk about some of these issues in my *Darwinism Defended: A Guide to the Evolution Controversies*, Reading, MA: Benjamin-Cummings, 1982.

83:29 An excellent book on the early history of life here on earth is Andrew Knoll, *Life on a Young Planet: The First Three Billion Years of Evolution on Earth*, Princeton, NJ: Princeton University Press, 2003.

84:1 Lynn Margulis, *The Origin of Eurkaryotic Cells*, New Haven, CT: Yale University Press, 1970.

85:32 *Life on a Young Planet*, 97.

87:5 M. Kimura, *The Neutral Theory of Molecular Evolution*, Cambridge: Cambridge University Press, 1983.

87:37 J. J. Sepkoski Jr., "A Kinetic Model of Phanerozoic Taxonomic Diversity. I Analysis of Marine Orders," *Paleobiology*, 1976, 4:223–251. I discuss Sepkoski's work in depth in *Mystery of Mysteries*.

90:9 Check out *Darwin and Its Discontents* for details.

90:37 P. Shipman, *The Man Who Found the Missing Link: Eugene Dubois and his Lifelong Quest to Prove Darwin Right*, Cambridge, MA: Harvard University Press, 2002.

92:9 There is a huge amount written on the Piltdown forgery. I touch on it in my *The Evolution Wars: A Guide to the Controversies*, Santa Barbara, CA: ABC-CLIO, 2000. My advice is to go to the Internet. Within minutes, you will find all that you need to know, and more.

92:29 A terrific read is D. Johanson and M. Edey, *Lucy: The Beginnings of Humankind*, New York: Simon and Schuster, 1981.

93:25 R. L. Cann, M. Stoneking, and A.C. Wilson, "Mitochondrial DNA and Human Evolution," *Nature*, 1987, 325:31–36.

95:3 P. Brown, T. Sutikna, M. J. Morwood, R. P. Soejono, Jatmiko, E. Wayhu Saptomo, and Rokus Awe Due, "A New Small-Bodied Hominin from the Late Pleistocene of Flores, Indonesia," *Nature*, 2004, 431:1055–1061.

97:3 I discuss the whole question of progress in biology, in much detail, in my *Monad to Man: The Concept of Progress in Evolutionary Biology*, Cambridge, MA: Harvard University Press, 1996.

97:20 P. B. Medawar, *The Art of the Soluble*, London: Methuen, 1967.

98:16 See, for example, Edward O. Wilson, *The Diversity of Life*, Cambridge, MA: Harvard University Press, 1992.

98:30 Stephen Jay Gould, "On Replacing the Idea of Progress with an Operational Notion of Directionality," in *Evolutionary Progress*, ed. M. Nitecki, Chicago: University of Chicago Press, 1988, 319–338; *Wonderful Life*, 318.

99:3 Stephen Jay Gould, "The Piltdown Conspiracy," *Natural History*, 1980, 89 (August): 8–28.

99:10 Dawkins discusses arms races in *The Blind Watchmaker*; Simon Conway Morris's ideas on progress can be found in his *Life's Solution: Inevitable Humans in a Lonely Universe*, Cambridge: Cambridge University Press, 2003.

99:25 Stephen Jay Gould, *Full House: The Spread of Excellence from Plato to Darwin*, New York: Paragon, 1996.

104:5 Cambridge, MA: Harvard University Press, 1975.[AQz3]

104:24 Cambridge, MA: Harvard University Press, 1978.[AQz4]

104:33 Michael Ruse and Edward O. Wilson, "The Evolution of Morality," *New Scientist*, 1985, 1478:108–128.

105:2　*On Human Nature*, 192.

105:23　See my *Sociobiology: Sense or Nonsense?* Dordrecht: Reidel, 1999.

106:12　Just so you don't think this is a fantasy, or itself motivated by anti-Semitism (Mayr was German born, although he came to America before the Nazi era), look at Stephen Jay Gould's discussion of the racially motivated laws in America in his *The Mismeasure of Man*, New York: Norton, 1981. I discussed this once with Gould and he agreed that there was something to it, although (surely, truly) he denied that this was his only or even the main motivation.

106:27　Martin Daly and Margo Wilson, *Homicide*, New York: De Gruyter, 1988.

107:14　An already-classic work is Brian Skyrms, *Evolution of the Social Contract*, Cambridge: Cambridge University Press, 1998.

107:32　Richard Dawkins, *The Selfish Gene*, Oxford: Oxford University Press, 1976.

109:33　An overview of the facts and theories can be found in my *Homosexuality: A Philosophical Inquiry*, Oxford: Blackwell, 1988.

111:29　Ronald Numbers in his *The Creationists* stresses just how much biological change is allowed by even the strictest literalist.

112:9　Genesis 9:24–27.

113:18　R. C. Lewontin, *Human Diversity*, New York: Scientific American Library, 1982.

113:12　R. C. Lewontin, *Biology as Ideology: The Doctrine of DNA*, Toronto: Anansi, 1991.

114:22　Leviticus 20:13; Romans 1:26–27.

115:1　Leviticus 19:19.

115:37　Sigmund Freud, "Letter to an American Mother," *American Journal of Psychiatry*, 1951, 107:787. The letter was written in 1935.

117:5　R. L. Trivers and D. E. Willard, "Natural Selection of Parental Ability to Vary the Sex Ratio of Offspring," *Science*, 1973, 179:90–92.

118:20　A lively discussion of this topic is Daniel Dennett, *Elbow Room*, Cambridge, MA: MIT Press, 1984.

119:6　Genesis 2:7.

119:13　I discuss this point in more detail in *Taking Darwin Seriously: A Naturalistic Approach to Philosophy*, 2nd ed., Buffalo, NY: Prometheus, 1998.

120:35　A. W. F. Edwards, "Human Genetic Diversity: Lewontin's Fallacy," *BioEssays*, 2003, 25:798–801.

121:10　He argues this in *On Human Nature*.

122:1　Stephen Pinker, *How the Mind Works*, New York: Norton, 1997, 138.

125:22　John Searle, "Minds, Brains, and Programs," *Behavioral and Brain Sciences*, 1980, 3:417–424.

127:28　Colin McGinn, *The Mysterious Flame: Conscious Minds in a Material World*, New York: Basic Books, 2000.

127:35　Richard Dawkins, *A Devil's Chaplain: Reflections on Hope, Lies, Science and Love*, Boston: Houghton Mifflin, 2003, 19.